上海城市空间艺术季展览画册编委会 | SUSAS PUBLICATION EDITORIAL BOARD

主编
CHIEF EDITORS

庄少勤 | ZHUANG SHAOQIN
胡劲军 | HU JINJUN
鲍炳章 | BAO BINGZHANG

副主编
DEPUTY CHIEF EDITORS

徐毅松 | XU YISONG
滕俊杰 | TENG JUNJIE
徐建 | XU JIAN

执行主编
EXECUTIVE CHIEF EDITORS

伍江 | WU JIANG
俞斯佳 | YU SIJIA
李翔宁 | LI XIANGNING

编委
EDITORIAL BOARD MEMBERS

(按姓氏笔画 | SORTED BY SURNAME STROKES)
王林 | WANG LIN
关也彤 | GUAN YETONG
李忠辉 | LI ZHONGHUI
沈竹楠 | SHEN ZHUNAN
张姿 | ZHANG ZI
张晴 | ZHANG QING
赵宝静 | ZHAO BAOJING
侯斌超 | HOU BINCHAO
奚文沁 | XI WENQIN
黄蕴菁 | HUANG YUNJING
章明 | ZHANG MING

2015
上海城市空间艺术季
案例展

Shanghai Urban Space Art Season
Site Project

主办：
上海市城市雕塑委员会
承办：
上海市规划和国土资源管理局
上海市文化广播影视管理局
上海市徐汇区人民政府

Host:
Shanghai Sculpture Committee
Organizer:
Shanghai Municipal Bureau of Planning and Land Resources
Shanghai Municipal Administration of Culture, Radio, Film & TV
People's Government of Xuhui District, Shanghai

上海城市空间艺术季展览画册编委会　编
Edited by SUSAS Publication Editorial Board

同济大学出版社
Tongji University Press

序言 Preface   1

上海城市更新的新探索 / 庄少勤
Exploration on Shanghai Urban Regeneration / Zhuang Shaoqin   3

城市更新与文化创新 / 胡劲军
Urban Regeneration and Cultural Innovation / Hu Jinjun   15

助力文化传承，勇于开拓创新，顺应城市工作新形势 / 鲍炳章
Boosting Cultural Heritage, Braving for Innovation and
Adapting to the New Situation of Urban Work / Bao Bingzhang   19

组织架构 Organization Structure   22

更 | 新场　浦东新场古镇实践案例展
Regeneration | New Place　Site Project of the Historic Town of Xinchang, Pudong   27

生活美学，点亮百年街巷　愚园路历史文化风貌区实践案例展
Lightening Time-Honored Street with Life Aesthetics　Site Project of Yuyuan Road Historic Area   51

对话　世博会城市最佳实践区实践案例展
Dialogue　Site Project of Expo Urban Best Practice Area in Huangpu District   67

小步伐，大改变　虹口区音乐谷实践案例展
Small Steps, Big Changes　Site Project of Shanghai Hongkou Music Valley   87

行走·跨越　上海天桥专题实践案例展
Walk · Cross　Shanghai Footbridge Thematic Exhibition   113

今天我们开向阳院　静安公共文化艺术实践案例展
Welcome to Our Sunflower Yard　Site Project of Public Art & Culture in Jing'an   133

CONTENTS
目 录

承上启下·见微知著　徐汇风貌区保护更新 2015 实践案例展
Transitional From and Subtle Gesture
Conservation and Regeneration in Xuhui Historic Area 2015 Site Project　　　145

空间漫步　"社区与技术"生活业态实践案例展
Space Tour　Site Project of "Community & Technology" Lifestyle　　　169

互动水乡　朱家角·尚都里实践案例展
Interactive Rivertown　Site Project of Zhujiajiao · Shangduli　　　191

重新装载　浦江东岸老白渡码头城市更新实践案例展
Reloading　Urban Renewal in Practice of Laobaidu Wharf, East Bank of Huangpu River　　　213

城市印记水岸传奇　闸北区苏河湾城市更新实践案例展
City Mark – Waterside Legend　Site Project of Zhabei District　　　229

宜居空间　普陀区曹杨新村实践案例展
Livable Space　Site Project of Caoyang New Village, Putuo District　　　245

寻·回　上楼下乡实践案例展
Seek · Return　Site Project of Rooftop Country Experience　　　259

上海城市规划展示馆系列展实践案例展
Serial Site Projects in Shanghai Urban Planning Exhibition Center　　　267

致谢 **Acknowledgements**　　　286

序言

上海市规划和国土资源管理局
上海市文化广播影视管理局
上海市徐汇区人民政府

上海经过 20 多年的快速发展，城市面貌焕然一新，市民生活明显改善，城市经济和活力令人瞩目。然而，与世界上很多大城市一样，由于人口加速集聚和城市快速扩张，环境、交通、安全等大都市的通病日益凸显，城市发展已经到了一个关键的历史性时期。上海城市面临的困难和挑战，城市未来的发展方向和发展方式迫切地需要得到社会各界的关注和积极参与，群策群力、众智众规。为此，上海市规划和国土资源管理局策划并发起两年一届的"上海城市空间艺术季"活动，希望打造一个以展示上海城市空间为平台的品牌活动，不断宣传和推广城市发展理念，促进城市空间塑造和公共艺术实践，展示城市优秀作品，呈现文化大都市魅力。

"上海城市空间艺术季"以"文化兴市，艺术建城"为理念。首届活动在上海市规划和国土资源管理局、上海市文化广播影视管理局、上海市徐汇区人民政府共同承办下，于 2015 年 9 月 29 日开幕，主题是"城市更新"，并通过"展览与实践"相结合的方式，将城市建设中的实践项目引入展览，将展览成果应用于城市建设中的实践项目。主展览设在徐汇滨江西岸艺术中心，通过"历史传承、创新畅想、乡村乡愁、公共艺术"四个角度诠释"城市更新"这一主题，将国内外先进的城市更新理论和实践全方位地进行展示和宣传。特别值得一提的是，不同于一般意义上的双年展，为加强实践性，本次空间艺术季活动在各区县设立了十五个实践案例展示，选择与百姓生活密切相关的公共空间，如传统街区、工业遗产、市政设施、绿化广场、社区空间等，通过国际策展与公开征集等多种方式，展示城市空间公共艺术改造方案及实施效果。同时，我们还将上海城市雕塑、公共艺术等相关展示活动及与市民生活密切相关的文化活动纳入形成联合展，使空间艺术季活动能不断地传播城市热点，美化城市空间，激发城市活力，提升城市魅力。

《2015 上海城市空间艺术季》一书是本届空间艺术季活动的重要成果之一。全书分上下两册，上册为主展览部分，包含主展览中"一条主线、四个维度、七个板块"的精华内容，下册为实践案例部分，包含所有十五个案例的空间创作和实践成果。这些丰富、翔实的内容，希望能为关心上海的城市更新工作的人们提供有益的借鉴，同时也期待更多市民共同参与到城市空间品质提升这一有意义的工作。

Preface

Shanghai Municipal Bureau of Planning and Land Resources

Shanghai Municipal Administration of Culture, Radio, Film & TV

People's Government of Xuhui District, Shanghai

Shanghai has taken on a new look with significantly improved life quality of its citizens and striking economy and vitality through more than 20 years of rapid development. Similar to other metropolises around the world, the urban development has entered a critical period when increasingly obvious problems appear in environment, traffic, safety, etc., due to rapid population increase and urban expansion. Attention and active engagement of all sectors of the society and wisdom and efforts of everyone are required to address the difficulties and challenges that Shanghai faces in determining urban development direction and pattern. To this end, Shanghai Planning and Land Resource Administration Bureau has planned and launched the biennial "Shanghai Urban Space Art Season" in the hope of building a brand event to showcase Shanghai urban spaces, communicate and promote the urban development concept, boost site projects in urban space shaping and public artworks, display excellent works in urban design and present the charm of the cultural metropolis.

Under the principle of "Culture Enriches City, Art Enlightens Space", Shanghai Urban Space Art Season (SUSAS) emphasis on "International, Public, and Practical", thus to improve the quality of public space and add the city with glamour. It hosted by the Shanghai Sculpture Committee and organized by the Shanghai Municipal Bureau of Planning and Land Resources, the Shanghai Municipal Administration of Culture, Radio, Film & TV and the People's Government of Xuhui District. This event played a key role in promoting the transformation of Shanghai, improving the quality of public space and boosting the urban regeneration. The event took "Expo + Site projects" model in order to promote a two-way exchange: introducing real cases from urban construction into the expo and applying the exhibits in practical projects. Located in West Bund Art Center in Xuhui Riverfront, The main exhibition interprets the theme of "Urban Regeneration" from four perspectives, namely "Historical Inheritance", "Innovation Creativities", "Rural Nostalgia", and "Public Art". Addition, different from the other biennales, the SUSAS put on fifteen real cases of urban public space most closely related to the lives of residents, such as traditional district, industrial heritage, municipal amenities, green space and square, community space and earth art with a focus on how art-led-regeneration can transform urban public space. Meanwhile, we will incorporate relevant shows of Shanghai urban sculptures, public arts, etc. and cultural activities closely related to the life of citizens to form extensive exhibitions that continuously spread urban hot topics, beautify urban spaces, and enhance vitality and charm of the city.

2015 Shanghai Urban Space Art Season is one of the important fruits of the event. The book consists of two volumes, one describing the main exhibitions and its essence of "one main line, four dimensions and seven sections", and the other presenting 15 site projects including space works and practical fruits. We hope that the rich and detailed contents will provide useful reference for people who are concerned about the urban renewal of Shanghai and are looking forward to more citizens' engagement in the meaningful work of improving the quality of urban spaces.

上海城市更新的新探索

在"世界城市日"学术研讨会上的讲话

庄少勤
上海城市规划委员会办公室主任
上海市规划和国土资源管理局 局长

城市更新伴随城市发展的全过程，折射生活方式和城市发展方式的变化，是城市持续发展和繁荣的驱动者。不同地区、不同阶段的城市更新，呈现出不同的特点。

上海从千年之前东海之滨的小渔村，到七百多年前江南水乡的新县城，从1843年开埠，到1990年的浦东开发，直至现在，一直处于城市更新的过程中。

与世界很多大城市一样，我们也经历了城市快速扩张、人口剧增等阶段，面临着旧城老化、服务能力不足等困扰。在这过程中，我们进行了以大规模旧区改造为代表的更新实践。但随着城市日趋长高、长大，以往"大拆大建"外延式扩张的发展老路已难以为继，注重提升城市品质和活力的内涵式发展成为当务之急。

I 上海城市更新进入了一个新阶段

挑战

上海的城市发展取得了举世瞩目的成就，但在土地利用、人口结构、空间品质、功能活力、文化传承及城市安全等方面也遇到了新的挑战。

一是土地利用。截止2014年底，全市建成区面积3124平方公里，超过市域陆地面积的45%，已逼近规划规模3226平方公里。与此同时，土地利用结构不够合理，工业用地比重过大，达27%。而公共设施和绿地的用地比例偏低。

二是人口结构。21世纪以来，上海人口激增800万，从2000年1608万人至2014年底已达到2425万人，人口老龄化、少子化特征日益明显。

三是功能活力。上海不仅在保持传统工业优势方面面临较大挑战，而且在互联网等新科技和新经济背景下，传统服务经济的发展也面临较大压力，创新经济发展任重道远。

四是空间品质。城市游憩空间不足，人均公共绿地7.1平方米。养老及社区文化、体育等公共服务设施也相对不足。

五是文化传承。以往以拆除重建为主的旧区改造方式，使上海城市历史风貌受到冲击。我们在延续历史文脉、留存城市乡愁方面要更加努力。

六是城市安全。上海也不时出现看海模式，全球气候变化异常等不确定性增加，对高密度、超大城市的安全和应急避难体系提出了更高要求。

目标

上海的国家使命是"当好改革开放的排头兵和创新发展的先行者",并要在建设"国际经济、金融、贸易、航运等四个中心"的基础上,建设具有全球影响力的科技创新中心。

上海在新一轮城市总体规划中的愿景是建设一座追求卓越的全球城市。着力从城市竞争力、可持续发展能力、城市魅力三个维度,打造更加开放的创新之城、更加绿色的生态之城、更加幸福的人文之城。

转型

2014年5月6日上海召开第六次规划土地工作会议,韩正书记明确提出了"上海规划建设用地规模要实现负增长",杨雄市长要求必须"通过土地利用方式转变来倒逼城市转型发展",这标志着上海进入了更加注重品质和活力的"逆生长"发展模式。

新的城市发展模式要求城市治理机制创新,必须探索一条城市更新的新路。着力在存量空间上,打造一座更有安全感、归属感、成就感和幸福感的全球城市。

II 上海城市更新的新概念

更新理念变迁

城市更新概念起源于西方,从"城市再开发"到"城市再生与城市复兴",现代城市规划理论则从"现代主义"到"新城市主义"再到"生态都市主义",甚至与东方的传统思想不谋而合。如1992年奥运会前,西班牙巴塞罗那借鉴中医理论,在城市更新中采取"针灸式"疗法,激发城市活力,成为东西方文化交融下的经典更新案例。

中国先贤们的世界观强调"天人合一",将城市看作天地间的有机生命体。在城市发展上遵循"以人为本,道法自然",在城市规划建设中提倡"有之以为利,无之以为用",讲究"因天材,就地利",强调城市整体性、协调性和持续性,是对空间系统的辩证的、有智慧的处理方法。其实,西方城市规划先驱帕特里克·盖迪斯(P.GEDDES)等学者也很早提出了必须将城市作为活的有机体的理念,不过被强大的工业化边缘化了。而经历过工业文明的洗礼后,东西方殊途同归,形成这样的共识:城市发展应当回归生态文明,回归到亚里士多德时代的理想——城市让生活更美好。

上海理念

上海的城市更新正是吸取了东西方文明的成果,形成了中西结合、具有上海特色的"城市有机更新"理念。

上海的城市有机更新不仅将城市作为有机生命体,也是将大城市作为若干"小城市"的共生群体;不仅将城市更新作为城市新陈代谢的成长过程,也将城市更新作为一种对城市短板的修补和社会的治理过程;不仅强调历史人文和自然生态的传承,也强调城市品质和功能的创造;不仅是城市发展质量和效益提升的过程,也是城市各方共建、共治、共享的过程。

上海特点

在城市"逆生长"的模式下,上海有机更新有以下特点:

一是更加关注空间重构与社区激活,把社区作为一个功能完备的"小城市"。构建以社区为基础"网络化、多中心、组团式、集约型"的城乡空间格局。

二是更加关注生活方式和公共空间品质,强调以人为本,围绕社区构建生活圈,增强公共空间的品质和人性化的场所体验。

三是更加关注功能复合与空间活力,改变工业文明机械式的区划分割的做法。适应创新经济时代需求,围绕人的创新创业活动,通过土地混合使用,打造功能合理复合的创新空间,激发城市产业活力。

四是更加关注历史传承与魅力塑造,突出城市特色,提升城市魅力,营造出兼具历史底蕴和现代气质的城市文化禀性。上海目前正在开展成片保护历史街区计划,如田子坊,单从建筑本身没有太大保护价值,但这个街区代表的城市肌理和市民的生活方式,是城市生命的有机组成部分。

五是更加关注公众参与和社会治理,城市更新不仅是空间重组过程,也是利益重新分配的过程。应发挥市民的主体作用,注重社会多元协同(包括规划者、建设者、运行者、管理者和需求者),构建和谐有序、共建、共治、共享的社会关系。

六是更加强调低影响与微治理,注重以"小规模、低影响、渐进式、适应性"为特征的"中医式疗法"更新方式,推动城市的内涵式创新发展。

III 上海城市更新的新方法和新实践

在学习总结多年实践经验基础上,上海市政府颁布《上海市城市更新实施办法》,强调上海实施动态、可持续的有机更新。并注重以下工作原则:

一是政府引导,规划引领,政府制定更新计划,以区域评估为抓手,落实整体更新的要求,发挥规划的引领作用。

二是注重品质,公共优先。坚持以人为本,以提升城市品质和功能为核心,优先保障公共要素,改善人居环境。

三是多方参与,共建共享,创新政策机制,引导多元主体共同参与,实现多方共赢。

四是依法规范,动态治理,以土地合同管理为平台,实施全要素全生命周期管理,确保更新目标的有效实现。

发挥文化引领作用,提升城市内涵

美国知名学者刘易斯·芒福德(L.MUMFORD)曾说"城市是文化的容器。城市根本功能在于文化积累、文化创新,在于流传文化,教育人民。"这段话精辟地阐述了文化、城市与人之间的关系。文化是城市的灵魂,发挥文化的引领作用,是对上海城市有机更新基本要求和首要任务。

在美丽的城市一定遇见美好的市民。其实,对市民而言,城市就是一个大众创作、大众享有的公共艺术品。市民不仅仅是城市文化的被动接受者,更是积极创造者。城市更新就是大众艺术创作。将艺术注入城市空间,用文化来引领城市更新,不仅可以提升城市品质,更可以提高市民品味乃至修养品行。正是基于这样的思考,我们以"文化兴市、艺术建城"为理念,在今年举办了首届"上海城市空间艺术季"。

不同于一般意义上的双年展,上海城市空间艺术季不仅面向所谓专业、精英人士,而是渗透到城市日常生活的每个角落,特别强调公众性和实践性。除主展览外,为进一步发挥街镇社区和市民作用,设立了十五个实践案例展,动态展示现实空间更新过程和效果。这也要求规划设计人员不仅要有人文情怀,还要有群众观念和实践能力。

发挥社区平台作用,完善生活圈功能品质

以社区为基本生活单元,打造生活圈。首先要以市民需求和社区文体

为导向，对更新地区进行综合评估，重点关注社区公共开放空间、公共服务设施、住房保障、产业功能、历史风貌保护、生态环境、慢行系统、城市基础设施和社区安全等方面内容，明确生活圈中"缺什么"，"补什么"，提供更加宜人的社区生活方式。

发挥市民主体作用，促进城市共享发展

坚持以民为本，保障市民权益。探索"政府—市场—市民—社团"四方协同的机制，注重物业权利人和设计师及政府部门的协作，发挥市民协商自治作用。努力避免将城市更新成为加重社会两极分化的过程。

发挥市场驱动作用，促进城市创新发展

上海的城市发展转型必须坚持减少对房地产的依赖。城市更新坚持公共利益优先，不以地块就地资金平衡为前提。一方面不能让市场太任性，否则容易产生不公等问题；另一方面要发挥好市场的资金、资源和创新能力等方面的作用。制定容积率奖励等方面的激励政策，多通过市场的方式，加强历史风貌保护、增加公共空间与改善公共服务。

以契约管理为抓手，创新城市治理机制

每一个更新项目实施就是一次城市治理行动。而持续有效的治理取决于参与主体持续、稳定的社会责任。通过对项目主体社会责任的全生命周期契约管理，如将物业持有比例等要求纳入合同管理，减少投机因素，使开发商转型为城市运营商，与城市共同成长。这样的社会契约关联可改善社会治理机制，从源头上减少城市病产生，促进城市共建、共治、共享。

IV 对未来上海城市的工作展望

城市更新已成为上海城市发展的主要方式，也是未来城市治理的关键抓手。我们将围绕"四个全面"的国家战略部署，主动适应和引领城市发展的新常态。根据"五位一体"的发展要求，不断探索"逆生长"模式下"有机更新"的新领域，持续提升城市品质和活力。

一是注重人本化。城市品质根本上取决于市民的品质，城市更新同样需要大众创新，通过人的更新成长来推动城市更新。我们正编制上海市民城市读本，办好 SEA-Hi 城市空间艺术跨界论坛，开展好"行走上海"、市民参与城市设计活动。

二是注重社会化。进一步完善社区规划体系与社区规划师制度，强化利益机制,如空间权益调节和激励机制。引导市场、社会主体和专业力量的积极参与,探索共享城市的治理模式。

三是注重信息化。积极应对智慧城市时代的要求,建立动态的城市体征指标评估体系,构建更加开放的网络化城市共享共治平台。

四是注重法治化。在进一步实践的基础上,适时提请市人大修订上海市城乡规划条例、上海市历史风貌保护条例,并进一步完善配套政策。

Exploration on Shanghai Urban Regeneration

Speech for Academic Seminar of World Cities Day

Zhuang Shaoqin
Office Director of Shanghai Urban Planning Committee and President of Shanghai Planning and Land Resource Administration Bureau

Urban regeneration accompanies the whole process of urban development, reflects changes in lifestyle and development mode of a city, drives the sustainable development and prosperity of a city. However, the features of urban regeneration vary from area to area, period to period.

From a humble fishing village by East Sea nearly one thousand years ago to a fledging town in Yangtze River Delta around seven hundred years ago, from the port opening in 1843 to Pudong Development in 1990, up to now, Shanghai never stops its regeneration.

With many of the metropolis around the world, Shanghai shares the experience of rapid expansion and population surge, as well as the problems of obsolete old towns and poor public services. We have practiced the regeneration featured by massive renovations. However, the city has grown out of the size for the old expansion and development mode featured by "mass-demolishing & mass-construction" and the connotation-based development that focuses on city quality and vigor improvement has grown high on the agenda for Shanghai.

I The Shanghai Urban Regeneration Has Entered a New Stage.

Challenges

Despite the glaring achievements, Shanghai still confronts new challenges on land exploitation, population structure, space quality, function and vitality, cultural heritage and city safety.

Firstly, land exploitation. By the end of 2014, Shanghai has 3,124 km^2 of built-up area, exceeding the 45% of urban or land area and approaching the plan area of 3,226 km^2. Moreover, Shanghai is plagued by an improper land exploitation structure, excessive industrial land proportion (27%) and insufficient coverage of public utilities and green land.

Secondly, demographic structure. Since the beginning of the century, the population in Shanghai surged from 16.08 million in 2000 to 24.25 million by the end of 2014, up by about 8 million, increasingly characterized by population aging and low birth rate.

Thirdly, function vitality. Confronting profound challenges on maintaining the edge in conventional industrial and the pressure to develop conventional economic services against the backdrop featured by such new technology and new economy as the Internet, Shanghai has have a long way to go to innovate economic development.

Fourthly, space quality. With only 7.1 m^2 of public green land per capita, the public recreational space for Shanghai is under-developed. So it is for the elderly caring, community culture and sport & public service facilities in Shanghai.

Fifthly, cultural inheritance. The previous renovation dominated by removal and reconstruction dents the historic charm of Shanghai. We should ramp up the conservation of cultural heritage and local nostalgia.

Sixthly, city safety. Stricken by water-logging from time to time, Shanghai, confronting increasingly changeable global climate, should call upon higher requirements for safe and emergency asylum systems in high-density metropolis.

Targets

The mission of Shanghai is to "pioneer the opening up and innovation development" and to become a technical innovation center with international influence based on the role of "international economic center, financial center, trade center and navigation center".

In the new overall urban planning, Shanghai has the vision to grow into an international metropolis striving for excellence and is committed to build a more open city of innovation, a greener environment-friendly city and a happier culture-oriented city by city competitiveness, sustainable development capacity and city charm.

Transformation

On May 6, 2014, Shanghai held the 6th Land Planning Meeting, where Han Zheng, General Secretary proposed to "de-grow the construction land in Shanghai" and Yang Xiong, Mayor of Shanghai requested to "transform Shanghai by transforming land exploitation". It marks the dawn of "de-growth" development mode of Shanghai, where attention is paid to quality and vitality.

As the new urban development mode requires for innovative urban governance mechanism, we should seek for a new approach for urban regeneration and strive to build an international city boasting stronger sense of safe, belonging, fulfillment and happiness.

II New Philosophy of Shanghai Urban Regeneration

Changes in regeneration philosophy

From "Urban Redevelopment" to "Urban Revitalization and Urban Rehabilitation", the idea of urban regeneration is originated from the West. Ranging from "Modernism", "New Urbanism" and to "Ecological Urbanism", modern urban planning theories even converge with Eastern traditional philosophies. For example, before 1992 Olympic Games, Barcelona, inspired by Chinese medicine theories, performed "an acupuncture therapy" to urban regeneration to vitalize the city, a classic case of urban regeneration that reflects the culture fusion between the West and the East.

The ancient sages of China upheld a world outlook that underlines "the Unity of Men and Nature" and regards a city as a living organism between the heaven and the earth. It is a dialectical and wise working method of space system to develop a city based on human-oriented approaches that follow the nature, resource availability and natural peculiarities and underline the globality, coordination and consistency of a city. In fact, a group of Western scholars, including Patrick Geddes, proposed the similar philosophy long time ago, i.e., to treat a city as a living organism. However, it was marginalized by the overwhelming trend of industrialization. Experienced industrial civilization, the West and the East join hands again and reach the consensus that urban development should return to ecological conservation and the dream dating back to the times of Aristotle: Better City, Better Life.

Shanghai Philosophy

Shanghai draws lessons from both the Western and Eastern cultures to formulate the "Organic Urban Regeneration" philosophy with Shanghai characteristics.

For organic regeneration of Shanghai, cities are living organisms, and each large city is a mutualistic symbiosis group. Urban regeneration is not only a metabolic process, but also a process to eliminate shortfalls and govern society. Focus should be paid to historic and cultural heritage and natural bestowment as well as city quality and function. The organic regeneration means not only enhancing urban development quality and efficiency, but also sharing construction, governance and resources.

Shanghai Characteristics

With the mode of urban "de-growth", the Shanghai organic regeneration has the following features:

1. More attention is paid to spatial reconstruction and community activation. Each community is regarded as a fully functional "small city". We will build a community-based urban-rural layout that is "network oriented, multi-centered, clustered and intensive".

2. More attention is paid to life style, public space quality and human-centered philosophy. Community-based life sphere is built to improve the public space quality and user experience of humane spaces.

3. More attention is paid to multi-functioning and spatial vitality to change the mechanical division characterized by industrial civilization. Adaption is made based on innovation and entrepreneurship to fulfill current economic requirements for creation, so as to build innovative spaces boasting rational and integrated functions and vitalize urban industries by mixed land exploitation.

4. More attention is paid to historic heritage and charm cultivation to highlight the characteristics and charm of Shanghai and develop a unique city identity that brings together historic heritage and modern radiance.

Shanghai is launching a plan to protect extensive historic blocks such as Tianzi Lane. Despite the limited protection value for buildings, this block represents the city texture and life style integral to Shanghai.

5. More attention is paid to public participation and social governance. Urban regeneration is not only spatial re-organization but an interest re-location. Citizens should play their major roles and focus on multilateral social coordination (including planners, constructors, operators, regulators and demanders) to create a harmonious, jointly built, corporately governed and shared society.

6. More attention is paid to minimize influence and subtle governance to achieve an urban regeneration characterized by "traditional Chinese medicine" thinking that boasts "small scale, minor influence, incremental progress and strong adaptation" to drive the connotative innovation development of Shanghai.

III New Methods and Practices for Shanghai Urban Regeneration

Based on years of practices and experience, Shanghai Municipal Government issued

Regulations on Implementation of Shanghai Urban Regeneration to emphasize that Shanghai would perform a dynamic and sustainable regeneration according to the following principles:

Firstly, government instruction and planned guidance. Resorting to regional evaluation, the government plans regeneration to address the requirements for overall regeneration and guide the work as planned.

Secondly, Quality counts and public first. Stick to the human-oriented approach and center on city quality and functions to safeguard public momentums and improve living environment.

Thirdly, multilateral participation and joint construction. Innovate policy-making system to guide multilateral participation and achieve win-win situation for stakeholders.

Fourthly, legal regulation and dynamic governance. Use land contract management as a platform to control whole life cycle for all factors to ensure effective fulfillment of regeneration targets.

Enhance the connotation of Shanghai by cultural guidance.

The famous US scholar Lewis Mumford, said "a city is the container of culture, of which the fundamental function is to accumulate culture, innovate culture, communicate culture and to educate people", which concisely illustrate the relation between culture, city and people. As culture is the soul of a city, culture development is a fundamental requirement and overriding task for the organic regeneration of Shanghai.

You will see nice citizens in a beautiful city. For local citizens, a city is an artwork created and shared by the public. Citizens are the creators of urban culture, rather than only receivers. Urban regeneration is the artwork created by the public. It helps to improve city quality, citizen tastes and accomplishment to inject art into urban spaces and guide urban regeneration with culture.

Therefore, we launched the first "Shanghai Urban Space Art Season" with the philosophy of "Rejuvenating a City with Culture and Building a City with Art" from September to December 2015.

Different from other biennales, Shanghai Urban Space Art Season addresses to the daily lives of common people in Shanghai rather than only those of the professionals and elites and emphasizes public participation and practice. Except for major exhibitions, 15 practice cases were presented to show how an actual space is regenerated and what it is ended up with, so as to give full play to the roles of local communities and citizens. It calls for the humanistic sensibilities, public care and practice capability of planners and designers.

Improve the living sphere with a community platform.

We build a living sphere based on communities, the fundamental living units. First of all, the area to be regenerated should be comprehensively evaluated based on citizens' needs and community culture. Special attention should be paid to such factors as the public open spaces, utility facilities, housing support, industrial functions, relic protection, ecology & environment, slow traffic systems, urban infrastructure and community security of a community to "define and offset the community shortfalls" and realize a more pleasant community life.

Citizens are taken as the main body to promote the shared development of Shanghai.

The people-oriented philosophy to protect the citizens' interests should be followed. We should develop the coordination of "government-market-citizens-associations" and underline the cooperation of property owners, designers and authorities to exert the autonomous negotiation of citizens. Escalation of social polarization caused by urban regeneration should be avoided.

The driving role of the market is fully played to promote the innovation of urban development.

The dependence on real estate industry of Shanghai urban transformation must be curbed. For urban regeneration, priority should be given to public interests rather than fund balance of the plot. We should control the market to avoid partiality and give scope to the fund, resource and creation of market. Such incentive policies as FAR awards should be formulated to improve historic heritage protection, public space coverage and public service by market.

Innovate the mechanism of urban governance by contract management.

Every regeneration project means an urban governance action. Persistent and effective governance depends on the persistent and reliable social commitment of major participants. Whole-life-cycle contract management on the social commitment of major project participants, i.e., to include the property holding ratio into contract management, can reduce speculation factors and turn developers into urban operators and make them grow with the city. Such social contract link helps to improve social governance mechanism and eradicate urban illness to enhance the sharing of construction, governance and resources.

IV Outlook for Shanghai Urban Regeneration

For Shanghai, urban regeneration has become a major growing approach and the key for future urban governance. Centered on the national strategic deployment of "Four Comprehensives" (comprehensively build a moderately prosperous society, comprehensively deepen reform, comprehensively implement the rule of law and comprehensively strengthen Party discipline), we should take initiative to adapt to and lead the new normal of urban development. Based on the requirements for "Five-in-one" development (economy, politics, culture, society and ecological conservation), we should keep exploring "Organic Regeneration" under "De-growth" mode and improving urban quality and vitality.

Firstly, people oriented: As the quality of a city depends on the quality of citizens, urban regeneration cannot do without public innovation. Urban regeneration should be promoted by the renewal and development of people. We are now formulating a city handbook for Shanghai citizens, running SEA-Hi forum, producing a documentary named "Walk in Shanghai" and encouraging citizens to participate urban design.

Secondly, society oriented: We improve the systems for community planning community planers and enhance interest control mechanism, such as space rights adjustment and incentive mechanism. Participation of market, social subjects and professionals were guided to explore the mode of shared urban governance.

Thirdly, information oriented: We address to the current requirements for smart city, build dynamic city evaluating index system and establish a more open network-based platform for joint building and corporate governance of city.

Fourthly, law oriented: Based on practices, we submit regulations of rural-urban planning and historic landscape protection of Shanghai to Municipal People's Congress for revision and improve the supporting policies.

城市更新
与文化创新

胡劲军
上海市文化广播影视管理局局长

万众瞩目的上海城市空间艺术季落下帷幕了。本届城市空间艺术季以"城市更新"为主题，聚焦"空间"和"艺术"关键词，充分演绎了城市更新的主题理念，展现了实践城市更新的经典案例，汇聚了探索城市发展的智慧思考，实现了市民参与城市文化氛围营造的诸多可能。

在城市加快更新的大背景下，上海城市发展既面临难得的发展机遇，也面临严峻挑战。究竟如何更好更快地发展和繁荣城市，提升市民生活、工作的城市公共空间品质，切实增强市民的生活幸福感和文化获得感？这需要城市更新与文化创造同步规划设计、同步组织推动。2015 上海城市空间艺术季的举办，正是谋求城市更新、城市发展的一次重要尝试。

这是一次以城市空间为载体，以文化艺术为内容，促进城市更新提升城市品质的创新实践。

文化是城市的灵魂和命脉。城市是文化、社会和经济活动的摇篮。文化的语言是城市最通俗的语言。上海城市空间艺术季将文化艺术作为城市更新的重要引擎内容，贯穿于活动的全程、城市更新的始终。

无论是演绎艺术季主题理念的主展览，展现城市更新多元实践的案例展、拓展艺术季内涵外延的联合展；还是激起智慧碰撞的主题论坛、激发参与热情的市民活动，艺术季以城市空间为载体，用文化活动激发了城市新的生机和活力，用艺术作品点亮了城市空间，营造了浓郁的城市公共文化氛围。

城市空间艺术季的举办，首次打造了具有"国际性、公众性、实践性"的城市空间艺术品牌活动，成为发挥城市文化创造力、凝聚力、影响力，更好服务城市更新发展、提升城市内涵品质的一次创新实践。

这是一次以跨界融合为引擎，以创造更新为路径，建立城市更新统筹协调平台的合力之举。

文化融合是世界文化发展的必然趋势。城市更新的过程是各种文化相互碰撞、激荡的过程，城市更新与文化创造的融合尤为必要。这有利于使各种文化和谐统一于城市更新中，形成合力，共同推动城市的整体发展。

本届城市空间艺术季以文化与规划的融合，文化、艺术、建筑、规划的跨界为特点，首次将艺术家、建筑师、策展人聚集在一起，创造了多个跨界融合的合力之举。在组织机构上，市区两级联动，文化与规划部门合作办节；在资源整合上，国际与国内专家联合策展；在平台联动上，上海城市空间艺术季与上海市民文化节首度联手，这一跨界融合、群策群力谋划城市更新的做法，在全国乃至全球范围内也尚属首次。

上海城市空间艺术季的举办，启动了跨界融合的引擎，吸引了众多跨界艺术家、多元社会主体，投入到上海城市更新和发展中来。并以文化创造与城市更新相结合为路径，为建立城市更新统筹协调平台，推进城市更新中文化的发展与繁荣提供了整体合力。

这是一次以城市更新为契机，以市民福祉为归宿，营造浓郁城市公共文化氛围的升级设计。

处于城市更新中，人们的思想活跃、观念多元、行为多向，这要求城市的发展，要特别注重把文化放到更为突出的位置。既重视城市更新中，营造浓郁的城市公共文化氛围，满足市民多样化的文化需求，又要注重以市民福祉为中心，增强市民的文化获得感和主体性。

城市需要文化，文化需要氛围。2015上海城市空间艺术季期间，透过上海市民文化节平台，持续推进美术、非遗、演艺、群文等各类展览、展示、展演，进地铁、进街头、进绿地、进商圈、进交通站点、进机场码头工程。在城市建筑、道路广场、轻轨车站、公园绿地等空间载体功能中，注入文化氛围因素，使文化无处不在、无处不见，广泛渗透于城市更新全过程。

城市是由生活其间的每个人构成的，市民不应该仅仅是文化的消费者、欣赏者，更应该是参与者、创造者和推进者。2015上海城市空间艺术季、上海市民文化节首度联手，共同推出"100个上海最美城市空间"和"100个上海城市空间塑造案例"的征集、评选活动。活动正是以市民为主体，邀请市民"为上海点赞""为城市支招"，激发更多市民参与到城市空间的塑造、创新中。以城市更新为契机，促进城市公共文化氛围营造的升级设计，推动上海城市空间更好地为城市居民服务。

城市空间艺术季的举办，是促进城市更新、提升城市品质的一次创新实践，是建立城市更新统筹协调平台的一次合力之举，是营造浓郁城市公共文化氛围的一次升级设计。这也仅仅是城市更新与文化创造有机融合，同步规划、合力共谋城市发展的一个起点。

Urban Regeneration and Cultural Innovation

Hu Jinjun
Director General of
Shanghai Municipal Administration
of Culture, Radio, Film & TV

The much-anticipated Shanghai Urban Space Art Season (SUSAS) has dropped its curtain. This Art Season is themed by 'urban regeneration' and focuses on 'space' and 'art'; the two key words fully express the thematic idea of urban regeneration, show typical cases of urban regeneration, gather wisdom and thoughts in regard of urban development, and realized citizens' participation in the city's culture construction.

In the process of urban regeneration , Shanghai is facing both opportunities and severe challenges. How to realize better and faster development of cities? How to improve the quality of urban living and working space, strengthen the citizens' life satisfaction and cultural acquisition? This requires a synchronized planning, design, organization and promotion of cultural innovation. The SUSAS 2015 is an important attempt in this direction.

This is an innovative practice based on urban space and focused on culture and art; it will promote urban regeneration and improve the quality of the city.

Culture is the soul and lifeline of a city. City is the cradle of cultural, social and economic activities. Cultural language is the plainest language of a city. SUSAS has taken culture and art as an important engine content of urban regeneration, and made it run through the whole activity and the urban regeneration process.

The main exhibition has conveyed the thematic concept of the Art Season; the case studies exhibition has showed diversified practices of urban regeneration; the joint exhibition has extended the connotation of the Art Season; the forum session has stirred up wisdom collision; and the citizen activities have inspired the citizens' enthusiasm in participation. Based on urban space, the Art Season has inspired the city's new vigor and vitality, lit up the urban space with art works, and created a rich urban public cultural atmosphere.

SUSAS has created the urban space art brand activities for the first time; the activities are international, public and practical. It has become an innovative practice that will utilize the creativity, cohesion and influence of urban culture as well as better serve the urban regeneration and development and improve the connotation quality of the city.

This is the fruit of cooperation; it is based on innovative renewal and uses cross-boundary fusion as the engine to establish the planning and coordination platform of urban regeneration.

Cultural integration is the inevitable development trend of world culture. The process of urban regeneration is a process that witnessed cultural collision and surge; that's why urban regeneration and fusion of cultural innovation is especially necessary. This will enable various kinds of cultures to unify harmoniously into the urban regeneration and form composition forces to jointly promote the overall development of the city.

SUSAS is characterized by the integration of culture and planning and the integration of culture, art, architecture and planning. It has gathered artists, architects and curators together to harvest the fruit of cooperation. In regard of organizational structure, the municipal level and the district level took joint action and the cultural departments and planning departments cooperated with each other; in regard of resources integration, international and domestic experts jointly curated the event; in regard of platform linkage, SUSAS cooperated with the Shanghai Citizen Cultural Festival for the first time. It's the first time in China even in the world to realize urban regeneration through such a cross-boundary fusion.

SUSAS has started the cross-border integration and attracted a lot of trans-boundary artists and pluralistic social main bodies to involve into the urban regeneration and development of Shanghai. It has also integrated cultural creation and urban regeneration to establish the planning and coordination platform of urban regeneration and provide overall joined forces to the cultural development and prosperity of urban regeneration.

We take urban regeneration as an opportunity to improve public welfare and create a richer cultural atmosphere in urban public space.

During the process of urban regeneration, people's thoughts are very active, concepts are diversified, conducts are multidirectional. This requires the urban development to focus on culture. When engaging in the urban regeneration, we should create a rich urban public cultural atmosphere, meet the diversified needs of citizens, and pay attention to public welfare and strengthen citizens' sense of cultural acquisition and ownership.

A city should create the right environment for culture development. During the SUSAS 2015, Shanghai continues to promote various kinds of exhibitions and demonstrations of arts, intangible cultural heritage resources, performing arts and folk arts. Relevant teams went to subways, streets, greenbelts, business circles, transport stations and airports and wharfs. The city has also injected cultural factors into spatial carriers such as urban buildings, roads, squares, light rail stations and park green spaces; this has enabled culture elements to be seen and felt everywhere and widely penetrated into the whole urban regeneration process.

A city relies on all the people living in it; citizens should not be merely consumers and appreciators, but should be participants, creators and impellers of culture activities. For the first time, SUSAS cooperated with the Shanghai Citizen Cultural Festival and jointly launched the solicitation and selecting activity of the "100 Most Beautiful City Spaces in Shanghai" and the "100 Shanghai Urban Space Establishment Cases". This campaign invited citizens to give their thumbs up for Shanghai and offer support, in this way inspired more people to participate in the building and improvement of the urban space. The Art Season has taken the urban regeneration as an opportunity to promote the upgrade design of the establishment of urban public cultural atmosphere and enable Shanghai's urban space to better serve the citizens.

SUSAS 2015 is a ceative practice to promote urban regeneration and improve the quality of the city; it is the fruit of cooperation to establish a coordination platform of urban regeneration; it is an upgrading design aiming at creating a rich urban public cultural atmosphere. What' more, it's also a starting point for the organic fusion of urban regeneration and cultural creation, synchronous planning and cooperated development.

助力文化传承，勇于开拓创新，顺应城市工作新形势

鲍炳章
徐汇区区长

上海城市发展已经进入新的阶段，如何借鉴发达国家城市的成功经验，改善城市人居环境，提高居民生活质量，丰富城市文化内涵，盘活土地存量而非增量，提升城市管理功能，实现城市公共服务设施、基础设施、文化设施的均衡供给，推动城市品质再上一个新阶梯，是上海城市发展面临的机遇与挑战，而城市有机更新将在文化创新的引领下成为应对这一机遇和挑战的有效方法。

徐汇区政府一直在不断地摸索和创新。在刚刚闭幕的上海城市空间艺术季中，主展览场馆就位于徐汇滨江的西岸艺术中心。对于徐汇滨江地区的开发，城市更新的重点在于土地的二次开发、用地性质和功能的转换、工业区转型以及滨水区的整治和改造，不"大拆大建"，一方面保留原有的工业建筑，加以更新改进成为保留城市工业记忆的新城区，例如工业塔吊、老船屋变身新的文化景观，一些航空油罐甚至成为极具创意的演艺空间，徐汇西岸艺术中心也是由上海飞机制造厂车间改造而成；另一方面，以现代文化艺术为背景打造"西岸文化走廊"品牌工程，着力提升城市公共开放空间品质，比如龙美术馆、余德耀美术馆、跑道公园等。同时，在滨江沿岸通过清除路障、加强安全配套设施、建立塑胶跑道等方式，艺术化地提升了街道空间品质，吸引了很多市民自主参与，推动了社区空间品质和活力的改善，把徐汇滨江打造为独具魅力的文化传媒产业集聚区和充满活力的滨水公共活动区。

除了主展览，本次艺术季中，我们还通过几个案例展——徐汇风貌区保护更新 2015、漕开发华鑫工业园区的实践案例展，以及从天桥展开的徐家汇城市更新市政特展，举办了内容丰富、形式多样的城市更新活动。这些活动将艺术注入到城市空间，用文化引领了城市更新。对于上海中心城区最大的风貌区——衡复历史文化风貌区，城市更新更应基于社区记忆和文化传承，着力弘扬海派城市文化。我们为此做出了许多探索性的保护更新措施，比如邀请陈丹燕女士做客城市更新论坛，口述社区记忆弘扬社区文化；举办尔冬强先生专题讲座，向公众解读 Art Deco 文化艺术；世界小学"老洋房探踪"特色拓展课程有力地显示了海派城市文化底蕴，为社区公共参与和建设提供了启示；风貌整治和老建筑改造设计交流研讨会从改进设计和实施模式等问题出发，分享了从建筑设计到管理和实施部门人员的交流，摸索出了相关切实有效历史风貌区建筑保护更新的做法；"徐汇风貌道路保护规划与实践 2007—2015"专业地展示了徐汇区在 2007 年以武康路为保护规划和整治实施试点到 2015 年全区风貌道路保护规划所取得的新思路和新模式。

在新的城市发展形势下，做好城市工作要抓住城市更新这条主线，继往开来，既要做好传统文化传承，又要勇于开拓创新，不断完善城市管理和服务，让居民在城市生活得更方便、更舒心、更美好。愿我们共同努力，建设好我们共同的城市。

Boosting Cultural Heritage, Braving for Innovation and Adapting to the New Situation of Urban Work

Bao Bingzhang
District Mayor of the People's Government of Xuhui District

As Shanghai has entered a new development stage, it is now faced with opportunities and challenges in how to use the successful experience of cities in developed countries to improve urban residential environment, enhance residents' living conditions, enrich urban cultural connotation, make good use of land reserves (rather than land supply), promote urban management functions, achieve balanced supply of public facilities, infrastructures and cultural facilities in the city as well as push forward the further advancement of urban quality; the organic regeneration of the city will be an efficient method to respond to those opportunities and challenges under the direction of cultural innovation.

The People's Government of Xuhui District has been in constant exploration and innovation. In the just-concluded Shanghai Urban Space Art Season, the main exhibition venue is in West Bund Art Center, which locates in riverside of Xuhui District. For the development of riverside area in Xuhui District, the focal point of city regeneration is the secondary development of land, the conversion of the nature and function of land, the transformation of industrial area and the renovation & retrofit of riverside area; demolition and reconstruction in large scale are to be prohibited, the original industrial buildings are kept and upgraded into the new area with the urban industrial memory on one hand such as the transfiguration of industrial tower cranes and old boathouses into new cultural landscapes; some aviation fuel tanks have even become innovative performance spaces. Xuhui District West Bund Art Center was renovated from the workshop of Shanghai Aircraft Manufacturing Factory. On the other hand, the branded project of "West Bund Culture Gallery" is to be built against the background of modern culture and art to improve the quality of urban public open spaces, such as Long Museum, Yuz Art Museum, Runway Park, etc. At the same time, the quality of street space has been improved artistically by clearing the roadblocks, enhancing the safety supporting facility, building plastic tracks, etc. along the riverside, a lot of citizens are attracted to participate voluntarily, in this way the improvement of quality and vigor of community space is boosted. The riverside area of Xuhui District has become a cultural media industry cluster district with unique charm and a waterfront of public activities that is full of energy.

During the Art Season, in addition to the main exhibition, we have also demonstrated a couple of case shows (cases in Xuhui Historic Area Regeneration 2015, Caohejing Development China Fortune Industrial Park, and special municipal exhibition of urban regeneration of Xuhui District that starts from overpasses) to hold urban regeneration events that is rich in various contents. Urban space is infused with art through these events, and the city regeneration has been led by culture. For the biggest historic area in downtown Shanghai - Hengfu Historic and Cultural Area, the urban regeneration shall be based further on the community memory and cultural heritage to propagate Shanghai urban culture. For this purpose we have taken a lot of explorative measures of protection and renewal, for example, we invited Chen Danyan to the City Regeneration Forum to tell about community memory and promote community culture; organized a special lecture for Er Dongqiang to interpret Art Deco culture and art to the public; Moreover, special curriculum Old Houses of World Primary School that vigorously demonstrated the cultural heritage of Shanghai urban culture and provided inspiration for the construction and participation of community public; feature renovation and design & reconstruction seminar derived from improvement design, implementation mode and other issues to share the experiences of personals from architectural design department to management and implementation department, developing a relevant and effective way for protection and renewal of historic area; Xuhui District Historic Road Conservation Planning and Practice 2007–2015 has demonstrated professionally the new ideas and new models acquired from the process from protective planning and implementation of renovation of Wukang Road as a trial in 2007 to the protective planning of the historic roads in the whole district in 2015.

In the new urban development situation, urban work shall follow closely the mainline of city regeneration; both the traditional cultural heritage and the bold exploration & innovation shall be emphasized; urban management and service shall be constantly improved to make city life more convenient, comfortable and happy. Let us work together to develop Shanghai into a better city.

组织架构
ORGANIZATION STRUCTURE

总策展人
CHIEF CURATORS

伍江 | WU JIANG
教授、博士生导师
同济大学副校长
法国建筑科学学院院士
Professor, Doctoral Supervisor
Vice-president of Tongji University
Fellow of French Architectural Academy of Science

莫森·莫斯塔法维 | MOHSEN MOSTAFAVI
建筑师、教育家
哈佛大学设计学院院长
Architect and Educator
Dean, GSD, Harvard University

执行团队
SUSAS EXECUTION TEAM

在上海市城市雕塑委员会统筹指导下推进艺术季具体工作开展
To work as coordinated and instructed by Shanghai Urban Sculpture Commission

俞斯佳 | YU SIJIA
上海市规划国土资源局总工
上海市城雕办副主任
Chief Engineer of Shanghai Municipal Bureau of Planning and Land Resources;
Vice Director of Shanghai Urban Sculpture Planning and Management Office

滕俊杰 | TENG JUNJIE
上海市文化广播影视局艺术总监
Art Director of Shanghai Municipal Administration of Culture, Radio, Film & TV)

徐建 | XU JIAN
徐汇区副区长
Vice District Mayor of People's Government of Xuhui District

上海市规划国土资源局风貌处
上海市文化广播影视局艺术处
上海市城市公共空间设计促进中心
徐汇区规土局
上海西岸开发（集团）有限公司
上海市公安局、市财政局、市建管委、
上海市绿化市容局、市旅游局相关处室
Landscape Division, Shanghai Municipal Bureau of Planning and Land Resources;
Art Division, Shanghai Municipal Administration of Culture, Radio, Film & TV;
Shanghai Design&Promotion Center for Urban Public Space;
Planning and Land Resource Administration Bureau of Xuhui District, Shanghai;
Shanghai Xi'an Development (Group) Co., Ltd; Shanghai Municipal Bureau of Public Security, Shanghai Municipal Finance Bureau, Shanghai Municipal Commission of Construction and Administration, Shanghai Municipal Greening Administration and Shanghai Municipal Tourism Administration.

学术委员会
SUSAS ACADEMIC COMMITTEE

组长 Leader

郑时龄 | ZHENG SHILING
中国科学院院士
Academician of Chinese Academy of Sciences

吴为山 | WU WEISHAN
中国美术馆馆长
Director of National Art Museum of China

成员
MEMBERS

按照姓氏笔画排序
Sorted based on surname strokes

马清运 | MA QINGYUN
美国南加州大学建筑学院院长
Dean of the USC School of Architecture

毛佳樑 | MAO JIALIANG
上海市规划协会会长
Chairman of Shanghai Municipal City Planning Administration

支文军 | ZHI WENJUN
原同济大学出版社社长
Former President of Tongji University Press

王澍 | WANG SHU
中国美术学院建筑学院院长
Dean of the School of Architectural Art, China Academy of Art

王建国 | WANG JIANGUO
东南大学建筑学院院长、城市规划研究院院长
Dean of School of Architecture, Southeast University and Urban Planning and Design School, Southeast University

王才强 | HENG CHYE KIANG
（新加坡 | SINGAPORE）
新加坡国立大学教授，环境与设计学院院长
Professor of National University of Singapore and Dean of School of Design & Environment NUS

北川弗兰 | FRAM KITAGAWA
（日本 | JAPAN）
国际策展人、大地艺术祭发起人
international curator and initiator of Echigo-Tsumari Art Triennial

伍江 | WU JIANG
同济大学副校长
Deputy President of Tongji University

朱子瑜 | ZHU ZIYU
中国城市规划设计研究院副总规划师
Deputy Chief Planner of China Academy of Urban Planning & Design

张永和 | YUNG HO CHANG
国家"千人计划"专家，同济大学教授
Member of the Recruitment Program of Global Experts and Professor of Tongji University

张杰 | ZHANG JIE
清华大学建筑学院副主任
Deputy Director of School of Architecture, Tsinghua University

李磊 | LI LEI
中华艺术宫副馆长
Deputy Director of China Art Museum

李向阳 | LI XIANGYANG
上海视觉艺术学院美术学院院长
Dean of School of Fine Art, Shanghai Institute of Visual Art

李振宇 | LI ZHENYU
同济大学建筑与城市规划学院院长
Dean of College of Architecture & Urban Planning, Tongji University

李晓峰 | LI XIAOFENG
上海大学艺术研究院副院长
Deputy Dean of Art Institution, Shanghai University

杨劲松 | YANG JINSONG
中国美术学院美术馆执行馆长
Executive Director of Museum of Contemporary Art of CAA

杨奇瑞 | YANG QIRUI
中国美术学院公共艺术学院院长
Dean of the College of Public Art, China Academy of Art

杨剑平 | YANG JIANPING
上海大学美术学院副院长
Deputy Dean of Academy of Arts, Shanghai University

汪大伟 | WANG DAWEI
上海大学美术学院院长
Dean of Academy of Arts, Shanghai University

沈迪 | SHEN DI
上海现代建筑设计集团副总经理兼总建筑师
Deputy General Manager and Chief Architect of Shanghai Xian Dai Architectural Design (Group) Co., Ltd.

玛莎·索恩 | MARTHA THORNE
（西班牙 | SPAIN）
普利兹克建筑奖执行理事
Executive Director of Pritzker Prize

芭芭拉·菲舍尔 | BARBARA FISCHER
（加拿大 | CANADA）
多伦多大学美术馆馆长
Director of University of Toronto Art Centre

尚辉 | SHANG HUI
全国城市雕塑建设指导委员会艺委会副主任
Deputy Director of Art Commission, Urban Sculpture Development Instruction Committee of China

郑佳矢 | ZHENG JIASHI
资深城市雕塑建设管理专家
Expert of Urban Sculpture Development and Management

郑培光 | ZHENG PEIGUANG
上海城市雕塑艺术中心副主任
Deputy Director of Shanghai Sculpture Space

施大畏 | SHI DAWEI
上海文联主席、中华艺术宫馆长
Chairman of Shanghai Federation of Literary and Art Circles and Director of China Art Museum

赵宝静 | ZHAO BAOJING
上海市城市规划研究院副院长
Deputy President of Shanghai Urban Planning & Design Institute

徐冰 | XU BING
中央美术学院教授
Professor of China Central Academy of Fine Arts

殷小烽 | YIN XIAOFENG
全国城市雕塑建设指导委员会艺委会副主任
Deputy Director of Art Commission, Urban Sculpture Development Instruction Committee of China

莫森·莫斯塔法维 | MOHSEN MOSTAFAVI
（美国 | USA）
哈佛大学设计学院院长
Dean of the Harvard Graduate School of Design

曹嘉明 | CAO JIAMING
上海市建筑学会理事长
President of the Architectural Society of Shanghai

曼纽尔·库德拉 | MANUEL KUDLA
（德国 | GERMANY）
国际建筑评论委员会秘书长，卡塞尔大学教授
General Secretary of International Committee of Architectural Critics and Professor of University of Kassel

隋建国 | SUI JIANGUO
国际著名艺术家
famous international artist

曾成钢 | ZENG CHENGGANG
中国雕塑学会会长
President of China Sculpture Institute

更 | 新场
浦东新场古镇实践案例展
Regeneration | New Place
Site Project of the Historic Town of Xinchang, Pudong

新场是一个不断生长的地方。

1300 年前，这里还是一片汪洋；800 年前，这里崛起了富甲一方的盐场；500 年前，这里遍布农桑兴盛的田园；300 年前，这里成为商贾辐辏的市集；100 年前，这里更是人文荟萃的名镇。

进入 21 世纪，城市化的波涛正激荡着曾经宁静的水乡。作为"上海浦东最后一块历史文化遗产地"（阮仪三教授语），新场古镇如何既能保护历经沧桑的风貌，又能汲取持续发展的活力？

紧邻上海自由贸易区和迪斯尼乐园，古老的社区面临全球性市场与文化的挑战。与此同时，当代中国齐头并进的"城市更新"与"乡土重建"运动又带来双重机遇。新场或者在传统商旅开发的侵蚀中名存实亡，或者通过自我创新的途径实现复兴。

古镇的更新是产业的更新，古镇的复兴是文化的复兴，古镇的发展是民众的发展。

开创一个承前启后的当代新场，既需要增强原住民对新兴市场的理解，也必须唤起外来资源对本地传统的尊重。仿佛脚下这片土地将在时代潮流的不断冲积中走向未来。

"新场古镇实践案例展"通过将艺术活动植入历史街区，引导传统文化融入当代生活。

遵循"上海城市空间艺术季"提出的城市更新主题，希望以展览为实验场，以艺术为催化剂，凝聚传统与时尚的精华，激发创意驻留新场，结晶出最美妙的新场实践。

新场，更新之地，更新之场。

Xinchang is a place of continual changes.

1,300 years ago, the place was still under the vast ocean; 800 years ago, a rich saltern was founded here; 500 years ago, the place became a cluster of thriving villages; 300 years ago, Xinchang turned into a market attracting numerous merchants; and 100 years ago, Xinchang was a well-known of great people and treasures.

In the 21st century, the trend of urbanization also greatly affects this tranquil waterside town. Since Xinchang is "the last historical and cultural relic at Pudong, Shanghai" (said by Professor RuanYisan), how to keep its historic features while maintain its vitality becomes a problem.

Located near Shanghai Free Trade Zone and Disneyland, the old community is facing challenges out of global markets and cultures. At the meantime, the "urban renewal" and "rural reconstruction" of modern China bring the old town double chances. Xinchang can only either exist in name only as a result of conventional development of commercial tourism, or rejuvenate itself by innovation.

Xinchang's renewal is industry-oriented, its rejuvenation is culture-oriented and its development is people-oriented.

To build Xinchang into a modern town that links the past and the future, we should not only enhance locals' understanding of emerging markets but also evoke foreigners' respect for local traditions. Just like the land we step on, driven by the wave of this new era, Xinchang will progress towards its future.

"Site Project of Xinchang Ancient Town" combines the artistic event with the historic street and introduces the traditional culture into the modern life.

On the theme of "Shanghai Urban Space Art Season", namely urban renewal, it is hoped that the exhibition may become a lab and art, the catalyst, so as to cohere traditional and fashionable factors, keep innovation in Xinchang and yield the most wonderful fruits.

Xinchang, as its Chinese name indicates, is a land and a place of renewal.

下图　更 | 新场第一展馆展示空间剖面图
BELOW Section of Exhibition Hall 1 of "Regeneration | New Place"

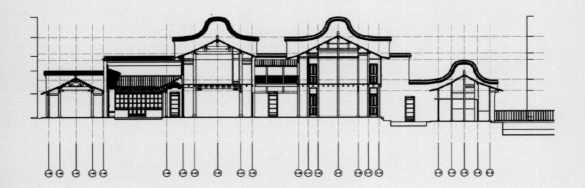

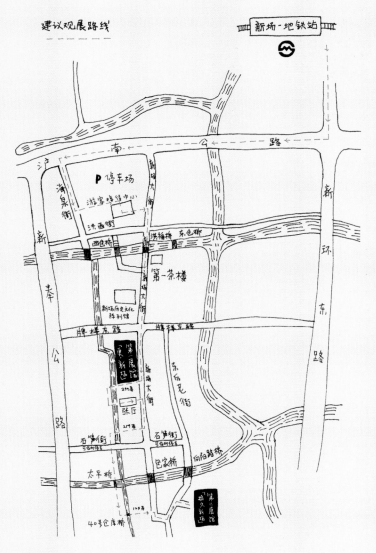

左图	新场建议观展路线
LEFT	Recommended visiting route of the 'New Place'

案例展展览主旨

"新场古镇实践案例展"通过将艺术活动植入历史街区,引导传统文化融入当代生活。鼓励有创业精神的青年一代关注新场丰富的文化遗产,通过多元化方式传承旧的手艺,酝酿新的文化精神。参展团队与艺术家从不同角度切入主题,创造互动体验,将沉寂多年的场地重新带入公共视野与街道生活之中。遵循"上海城市空间艺术季"提出的城市更新主题,希望以展览为实验场,以艺术为催化剂,凝聚传统与时尚的原料,激发创意驻留新场,结晶出最美妙的新场实践。新场,更新之地,更新之场。

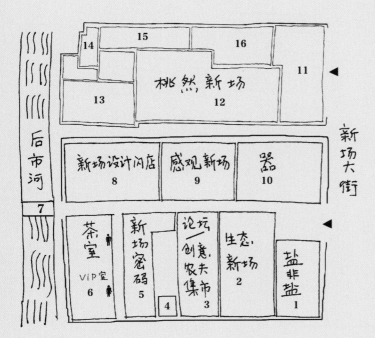
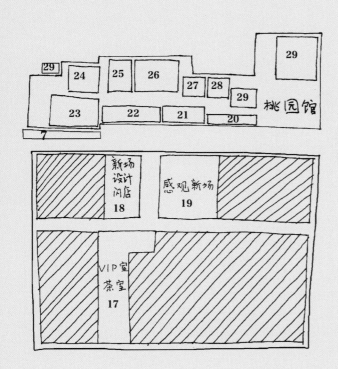

左图　更｜新场第一展馆平面图
LEFT　Plan of "Regeneration｜New Place" Exhibition Hall 1

1	盐非盐 盐雕 "More Than Salt", Salt Carving	11	桃然新场 手作弄堂 Taoran & Xinchang, Hand-made Alley	21	更新场景 在地感应装置 "Updated Scene", on-site sensing device
2	生态新场 生态装置 "Ecological Xinchang", Ecology Device	12	桃然新场 酵堂 Taoran & Xinchang, Ferment Class	22	生成 摄影装置 "Generate", Photography devices
3	论坛 / 创意农夫集市 Forum/Creative Farmer Market	13	桃然新场 讲堂 Taoran & Xinchang, Lecture Hall	23	虚拟与现实 多媒体艺术实验室 "Virtual and Reality", Multimedia Art Laboratory
4	汽车时代 空气雕塑 "Auto Era", Air Sculpture	14	桃然新场 桃园 Taoran & Xinchang, Peach orchard	24	吃火吐水 行为影像 "Fire & Water", Performance video
5	新场密码 展览互动 "Xinchang Password", Interactive exhibition	15	桃然新场 食堂 Taoran & Xinchang, Dining Hall	25	百戏 Bai Xi
6	VIP 茶室 当代装置 "VIP Tea House", Contemporary Device	16	桃然新场 灿计划 Taoran & Xinchang, C.A.N.	26	甚至怀有恐惧 装置 "Even fear", Installation
7	花落花开又一春 在地生命装置 "Flowers Fallen and Another Spring", on-site life installation	17	VIP 茶室 当代装置 "VIP Tea House", Contemporary Device	27	银河 当代雕塑 "Galaxy", Contemporary Sculpture
8	新场设计闪店 新场建筑空间主题展 "Xinchang Design Flash Store", Xinchang building space theme exhibition	18	新场设计闪店 文创展示 "Xinchang Design Flash Store", Cultural & Creative exhibition	28	山河水 Mountain River
9	感观新场 多媒体互动 "Sensory Xinchang", Interactive Multimedia	19	感观新场 新场文化体验展 "Sensory Xinchang", New Place Cultural Experience Exhibition	29	光雕塑 灯光装置 "Light Sculpture", Light Installation
10	器 茶、陶艺展示 Tea & Pottery show	20	化妆风景 影像作品 "Makeup Landscape", Image work		

展览介绍

I 第一展馆: "更·新场"
鼓励有创业精神的青年一代关注新场丰富的文化遗产，通过多元化方式传承旧的手艺，酝酿新的品牌。参展团队与艺术家从不同角度切入主题，创造互动体验，将沉寂多年的场地重新带入公共视野与街道生活之中，使古镇中的老宅焕发新生。

1. 感观新场
展览意在将新场古镇的空间文化置入古镇生活、传统风貌、民俗活动之中，从不同角度探究水乡古镇场所文化，融合多重感官体验，以多媒体的交互为手段，提供可观、可感、可触、可听、可思的多维展示空间，探索新场古镇的历史、现在和未来。

上图	更｜新场第一展馆展示空间——《茶室》
ABOVE	Exhibition Hall 1 of "Regeneration｜New Place"—"Teahouse"

左图	花落花开又一春
LEFT	"Flowers Fallen and Another Spring"

"感观新场"一层：更·新场—新场建筑空间主题展示；"感观新场"二层：感·观·新场文化体验展。

本展区是"更新场"展览的主展场，共分上下两层。展览意在将新场古镇的空间文化置入古镇生活、传统风貌、民俗活动之中，从不同角度探究水乡古镇场所文化，融合多重感官体验，以多媒体的交互为手段，提供可观、可感、可触、可听、可思的多维展示空间，探索新场古镇的历史、现在和未来。

本页　更｜新场第一展馆展示空间
THIS PAGE　Exhibition Hall 1 of "Regeneration | New Place"

2. 器

自古以来，茶楼书馆就是新场民间文化与生活的中心。以"器"为线索的展览，由当代人重新诠释"瓷、茶、琴、书、画"等传统艺术活动，将古镇建筑空间与传统文化艺术、物质与非物质文化相融，达到激活历史文脉，传承传统文化的双重效果。作品提取古镇建筑特色及相关纹饰元素，为古镇限量定制出茶具产品。同时创作出以代表新场历史的《敖波图》为意象、可供游客参与手工制作的展示台，从而让游客亲身体验古镇独特的文脉。

3. 新场设计闪店

"新场设计闪店"是一个有关生活与艺术的概念性展示空间,也是一次远离都市的奇妙体验。场地位于上海浦东新场古镇内一处有百年历史的民宅,参加展览的九位艺术家分别来自上海,杭州,北京,共同进驻为期两个月的"新场设计闪店",并围绕"Take Time"的主题,用各自的作品共同构建一幅融合传统与幻想的生活场景。展场设计的灵感来自江南水乡的建筑及手工艺材料。仿佛生活本身,展览不会一成不变,新与旧的事物相互交替实现动态的平衡。

本页 "新场设计闪店"中的作品
THIS PAGE Works from "Xinchang Innovative Market"

4. 盐非盐

盐是生活味道的来源，也是新场建镇的初始。以"盐非盐"为主题的展示与互动装置，意在感悟和发掘盐的文化艺术内涵和品德，用当代艺术的语言表达盐的视觉艺术力量。盐是盐，是超越时代的生命基础；盐又非盐，代表传统社会中勤劳和智慧的精华。盐的特性可以反腐保鲜，作品以此寓意新场古镇的活性延续。

5. 新场密码

"新场密码"是一个关于新场的探索游戏。活动以新场老建筑的经典细节为线索，以复原新场生活场景为媒介，呈现手工业时代的匠人用心。策展人挑选20～30栋新场经典老建筑，对其中的建筑细节进行提炼，通过视觉影像化的转译手法，将它们制作成虚拟性的建筑密码，并集中显现于"更新场"的主展厅。作品希望引导参观者深入发现新场的潜在价值，并使用新媒体技术呈现这种激动的体验。

1、2　　盐非盐
　　　　More Than Salt
3、4、5、6、7　新场密码
　　　　Decode History of Xinchang

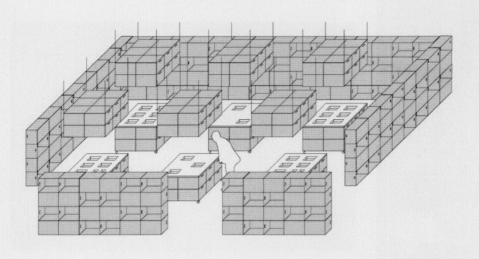

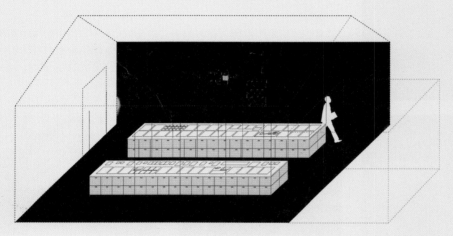

6. 桃然新场

"桃然新场"由南加州大学建筑学院院长马清运教授策划,希望在新场建立一个展览、学术讨论和社区活动的平台。一方面,让世界了解中国农业文明的传统与现状、问题与机会。另一方面,在展现"灿计划"实践项目经验的基础上,让更多专业和金融人士参与讨论,一起探寻 C.A.N. 的解决之道——Agri-urbanism(农业都市主义)。"灿计划"(C.A.N.)旨在探索文明与自然之间的共生机制。英文 C.A.N. 是 Culture, Agriculture and Nature 的缩写。继威尼斯双年展和包豪斯论坛之后,"灿计划"落地新场,与本地社群共同讨论"未来古镇＋农村发展"的相关重要话题。这次受邀参与"更｜新场"展览的"灿计划"一共分 5 个部分,分别为:C.A.N.、酵堂、弄堂、食堂、讲堂。

对页及本页 桃然新场现场展示空间
OPPOSITE & THIS PAGE Exhibition Hall of "Taoran & Xinchang"

桃然新场现场展示空间
Exhibition Hall of "Taoran & Xinchang"

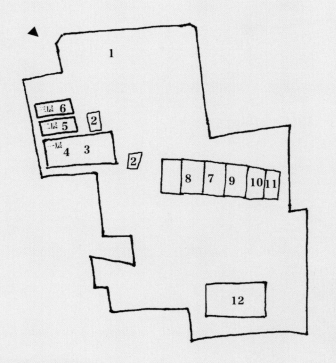

左图	第二展馆"拾久新场"（原针织十九厂）平面图
LEFT	Plan of second exhibition hall, Memories of Xinchang (No. 19 Shanghai Knitwear Mill)
1	灯光装置 Light installation
2	宋应星的雕塑 装置 "Song Yingxing's Sculpture", installation
3	山水余韵 综合装置 "The remain of the beautiful landscape", Comprehensive installation
4	山水余韵 综合装置 "The remain of the beautiful landscape", Comprehensive installation
5	皮肤 综合装置 "Skin", Comprehensive installation
6	空间叙事 绘画 "Space narrative", Painting
7	四十万公里 影像装置 "Four hundred thousand kilometres", Video installation
8	太极 雕塑 "Tai Ji", Sculpture
9	快砖项目 复合形式 "Fast brick combination", Composite form
10	色彩趋势 调研报告 Color Trend, Investigation Report
11	日用工业 声音装置 "The Daily Industry", Sound installation
12	花镜屋楼 灯光装置 "Mirage glasses", Light installation

II 第二展馆："拾久新场"

"拾久新场"作为更｜新场的第二展馆，突破了常规传统展馆的物理空间，而是在新场古镇的一个区域置入艺术品。24件艺术品或室内或室外的与新场的建筑、公共空间融合，在新场这一场域沉淀、发酵，使区域焕发出新的生机。

拾久新场的室外装置作品《宋应星的雕塑》是胡项城、李晓捷、宋娇合作的作品，宋应星是明代著作《天工开物》的作者，书中记录了中国农耕时代的产业与生活有关的发明创造。在这组装置中高高叠立上方的农具，象征了一代代先人的勤劳智慧的结晶。艺术家们思考如何将这些本可使用的"雕塑"再次在农田应用，可以通过影视记录，使器物与应用方式保存给一代又一代。世事变化无常，任何先人的经验，不知何时也许又会有用武之地。

另一个室外作品《花镜屋楼》是一个灯光装置，由伊天夫、沈倩和十余名艺术工作人员参与设计制作。作品由两部分组成，第一部分是在十九厂室外空间装置99面镜子与特殊日光灯及各种光源组成奇异的景观，镜中反射的月光与人工光源共同反映了人类对外太空世界及未来的出神片刻。第二部分装置是用传统木构架与宫灯组成，9月27日是中秋

《宋应星的雕塑》装置 宋娇（胡项城指导合作）
"Song Yingxing's sculpture", installation. Artist: Song Jiao; Tutor: Hu Xiangcheng

1	2	6
3	4	
5		7

1 花镜屢楼
  Mirage glasses

2 皮肤
  Skin

3 空间叙事
  Space narrative

4 阿米种子图书馆
  Amie seed library

5 四十万公里
  Four hundred thousand kilometres

6 快砖组合
  Fast brick combination

7 近于禅
  Close to zen

佳节，这组装置使新场古镇的传统景观得到延伸展现。传统与现代两组景观同时出现在废弃的新场镇十九厂，加入音乐多媒体影像，体现了中国天人合一的宇宙观与生生不息的创新精神。

《四十万公里》是艺术家周长勇的作品。这件作品是世界上第一件使用 LED 灯光制作的三维矩阵效果的动态雕塑。它是由上万个 LED 点光源组成。在高 2.5 米，宽 80 厘米，长 100 厘米的三维矩阵空间展示一个人在不停地行走状态。

《快砖组合》是一组跨媒体装置，该作品由闫雪、宋娇、刘一博、卢奕达、王晓霖、李晓捷、李赏悦、郑湘竹八位年轻人以自己不同的方式来表达对传统与现代、个体与社会等多方面的看法。这组作品用了最司空见惯的建筑的材料——砖，寓意每一个人是构成社会的一砖一瓦。当然即使最新的砖也会成为未来废墟的残片，但正是这一层层的累积和这无数的临时才形成了人类文明的大厦。

Theme of Case Exhibition

"Site Project of Xinchang Ancient Town" combines the artistic event with the historic street and introduces the traditional culture into the modern life. The aim is to encourage young entrepreneurs to pay attention to the rich cultural relics of Xinchang, to pass on the traditional handicrafts in diversified ways and to brew new culture and spiit. Touching the theme from different aspects, relevant groups and artists have created many interactive artworks to bring this long-forgotten land back into people's mind and daily life. On the theme of "Shanghai Urban Space Art Season", namely urban renewal, it is hoped that the exhibition may become a lab and art, the catalyst, so as to cohere traditional and fashionable factors, keep innovation in Xinchang and yield the most wonderful fruits. Xinchang, as its Chinese name indicates, is a land and a place of renewal.

Exhibition Introduction

l Exhibition Hall 1:" Renewal • Xinchang"

The aim is to stimulate young entrepreneurs to pay attention to the rich cultural relics of Xinchang, to pass on the traditional handicrafts in diversified ways and to develop a new brand. Touching the theme from different aspects, relevant groups and artists have created many interactive artworks to bring this long-forgotten land back into people's mind and daily life and to rejuvenate old houses in this ancient town.

1. Feeling Xinchang

This exhibition intends to demonstrate Xinchang's space and culture with its ancient town life, its traditions and its custom, so as to explore the culture of this ancient waterside town from different aspects and with different senses. Meanwhile, with multi-media interactive technology, a comprehensive exhibition that can be seen, felt, touched, heard and considered has been formed so that one can explore Xinchang's past, present and future.

"Feeling Xinchang" F1: "Renewal · Xinchang" - Xinchang Architectural Space Exhibition is held here. Feeling Xinchang F2: "Feeling Xinchang" -Cultural Experience Show is held here.

This 2-storey building is the main exhibition pavilion of "Renewal · Xinchang" site project. This exhibition intends to demonstrate Xinchang's space and culture with its ancient town life, its traditions and its custom, so as to explore the culture of this ancient waterside town from different aspects and with different senses. Meanwhile, with multi-media interactive technology, a comprehensive exhibition that can be seen, felt, touched, heard and considered has been formed so that one can explore Xinchang's past, present and future.

2. Tea Ware

Since the ancient time, tea houses and story-telling houses have been the center of people's life and culture in Xinchang. "Tea ware" being the clue, this program demonstrated traditional artistic actions in a new way, including those related to "porcelain, tea, music, calligraphy and painting". By combining the architectural space and the traditional culture and art of this old town, as well as material culture and intangible culture, the program aims at both stimulating and passing on the historic culture. The artwork was designed on the basis of classic structures and relevant ornamentation of this old town. These were limited tea ware especially customized for this town. At the meantime, the program also included a show stand based on the image of The Picture of Boiling Sea Water for Salt. Visitors could enjoy the happiness of DIY there and experience the unique culture of this old town.

3. Xinchang Innovative Market

The "Xinchang Innovative Market" is a conceptual exhibition hall demonstrating life and art. Visitors can enjoy a wondrous tour here away from metropolitan influences. The market is located at a century-old house at Xinchang, Pudong District, Shanghai. Nine artists from Shanghai, Hangzhou and Beijing would run their stalls at the "Xinchang Innovative Market" on the theme of "Take Time" for two months. At this market, these artists have co-built a series of scenes featuring tradition and fantasy with their artworks. The designers have been inspired by the architectures and handicrafts in Jiangnan Region. Similar to life, the exhibition does not remain unchanged. In fact, the old and the new would keep replacing each other to form a dynamic balance.

4. More Than Salt

Salt is the source of taste in our daily life and XinchangTown was also established on salt. The purpose of this demonstrativeand interactive installationthemed "More Than Salt" is to help people discover and understand the culture, artistic properties and virtues of salt, and to show the visual impact of salt by means of contemporary art. Salt is salt for it's the basis of life at all times; however, salt is more than salt for it represents the essence of diligence and wisdom of ancient Chinese. Because salt can prevent decay and keep things fresh, this installation implies that the vitality ofXinchang Ancient Town will last.

5. Decode History of Xinchang

"Decode History of Xinchang" is an adventure about Xinchang. With classic details of old buildings at Xinchang as the clues and reconstructed scenes as the media, the exhibition shows the dedication of craftsmen at that era of handicraft. The curator has selected 20 to 30 classic ancient buildings at Xinchang and extracted architectural details from these buildings. The details have then been visualized into virtual architectural codes displayed at the main exhibition pavilion of "Renewal · Xinchang". The artwork aims to help visitors understand the potential of Xinchang in depth. To achieve this goal, the exciting new media technology is adopted.

6. Taoran & Xinchang

"Taoran & Xinchang" is a project run by Professor Ma Qingyun, Dean of the School of Architecture of the University of Southern California. Professor Ma intended to build a platform for exhibition, academic discussion and community events at Xinchang. On one hand, the project may help the world to understand the history, status quo, problems and opportunities of Chinese agricultural civilization. On the other hand, while showing the experience gained during C.A.N. project, the site project can attract more professionals and financial figures into the discussion and exploration of a C.A.N. solution, namely, Agri-urbanism. The purpose of "C.A.N." project is to find out a symbiosis mechanism of civilization and nature. "C.A.N." is the abbreviation of Culture, Agriculture and Nature.

After its debut at Venice Biennale and Bauhaus Forum, "C.A.N." set foot at Xinchang and, along with local communities, became a part of the crucial discussion about "Future Ancient Town + Rural Development". "C.A.N." project at the "Renewal · Xinchang" site projectconsisted of five parts: C.A.N., the fermentation room, the alley, the eatery and the classroom.

‖ Exhibition Hall 2: Memories of Xinchang

At "Memories of Xinchang", the second "pavilion" of Regeneration|Xinchang, artworks are placed in one area of Xinchang, without the spatial limits on traditional pavilions. This area is invigorated with 24 indoor and outdoor artworks integrated in public areas and the buildings and to be mellowed as time goes on.

The outdoor work "Song Yingxing's Sculpture" is built by Hu Xiangcheng, Li Xiaojie and Song Jiao together. Song Yingxing was the author of Tiangong Kaiwu (literally, exploitation of the works of nature) from the Ming Dynasty, which recorded inventions and creations related to industries and daily life in the Chinese Agrarian Age. The agricultural implement standing up high in this device represents the hard-work and wisdom of ancestors of the Chinese people. The artists are thinking about using this "sculpture" in the farmland and shooting the scene to keep an image of both the implement and its application for next generations. As the times change, maybe someday heritage from the ancestors will fit in again.

"Mirage glasses", another outdoor work, is a lighting device designed by Yi Tianfu, Shen Qian and a dozen more. This work consists of two parts. For the first, 99 mirrors are installed outdoor, complete with special fluorescent lamps and a variety of light sources to form a spectacle where the moonlight reflected from the mirrors and the artificial lighting together represent the spellbound moments of human beings looking into the outer space and the future. The second part is made of traditional wooden frames with palace lanterns. September 27th is the Mid-Autumn Festival, and the device will be an extension of the traditional scenery of the ancient town. The old and the new, the two installations in the abandoned Nineteen New Place, complemented by music and multimedia images, represent the Chinese theory that man is an integral part of nature and the growing spirit of innovation.

"Four hundred thousand kilometres", by Zhou Changyong, is the world's first three-dimensional dynamic matrix sculpture made of over ten thousand LEDs. The artwork presents a man walking in a three-dimensional matrix space, which is 2.5 m high, 80 cm wide and 100 cm long.

"Fast brick combination" is a set of cross-media device. Eight young artists Yan Xue, Song Jiao, Liu Yibo, Lu Yida, Wang Xiaolin, Li Xiaojie, Li Shangyue and Zheng Xiangzhu express by this device their ideas about the relationship between traditional and modern, individual and the society. The most common material bricks are used, meaning everyone is a brick in the community. Even a brand-new brick will become a fragment of ruins in the future, but the mansion of human civilization is built up with these layers and numerous temporary moments.

策展人感想

王林

城市空间作为城市的一个生命细胞，空间品质的提升与活力的激发是推动上海城市有机更新的重要力量。上海未来的发展将把品质作为核心要素，尤其是城市空间品质。其提升与完善除了需要创新和地方营造，发挥地域资源优势，开发传统文化资源，更要注重文化建设和上海中心城风貌及周边古镇历史脉络的保护与传承。2015上海城市空间艺术季新场古镇实践案例展就是一次在历史风貌区提升空间品质、激发文化活力的实践探索。借力艺术活动植入历史街区，引导传统文化融入当代生活，将沉寂的场地重新带入公众的视野和社区生活，激励古镇空间的更新和生长。

胡项城

曾经的上海针织十九厂坐落在新场古镇。上海空间艺术季新场分展场将在这废弃已久的厂房开始"拾久新场"创意园的正式启动。来自各地的艺术家与本镇的群众近50余人，在这里用当代诗画、当代雕塑、影像、装置、光雕塑、生命装置、民俗表演等展示人类各种生存空间的今昔与未来的探求。这次展览突破了封闭在专业场馆的限制，真正走向社会的核心——民间。艺术不再是清高无上的标志，人与人之间需要平等沟通的心理与物理空间。在两个月的展示期间，这里还将举行各种活动，包括艺术与社会等内容的研讨。

Note from the curator

Wang Lin

City space as a living cell of the city, improve the quality of space and energy excitation is an important force to promote the Shanghai city organic renewal. The future development of Shanghai will put the quality as the core elements, especially the city space quality. Its improvement and perfection in addition to local innovation and create, play the advantages of local resources, the development of traditional culture resources, we should pay more attention to the protection and inheritance of the Xinchang Town Exhibition at SUSAS 2015 is a practical case in the historic district space to improve the quality of exploration and Practice to stimulate the cultural vitality. Leveraging the arts implanted Historic District, to guide the traditional culture into modern life. The silence of the site back into public view and community life, growth and regeneration of incentive town space.

Hu Xiangcheng

No. 19 Shanghai Knitwear Mill was located in Xinchang. At this long deserted factory, the commencement ceremony of "Xinchang in No.19 Mill" innovation park of Xinchang District of Shanghai Urban Space Art Season was held. Over 50 people, including artists from different regions and local citizens, demonstrated the past and the present as well as the possible future of various living spaces with modern poems, modern sculptures, video clips, instruments, light sculptures, live units and folkway shows. This program broke the limit of exhibition in closed professional exhibition halls and truly touched the core of the society — ordinary people. Art was no longer a distant symbol. People needed a physical and mental space for equal communication. During the exhibition which lasted for two months, the factory would witness various events, including seminars on art and society.

策展人 CURATOR

王林
上海交通大学建筑系教授，城市规划博士，哈佛大学研究学者。同济大学兼职教授，上海市规划委员会专家委员，国家历史名城保护学术委员会委员，城市土地学会委员。

Wang Lin
Professor of Department of Architecture in Shanghai Jiao Tong University, PhD in Urban Planning, Harvard University Visting Scholar. Adjunct professor in Tongji University; Expert member of Shanghai Municipal Commission of Urban Planning; Member of National Historic City Conservation Academic Committee; Member of Urban Land Institute.

胡项城
著名当代艺术家，曾任教于上海戏剧学院、西藏大学，上海双年展创立者之一，曾任上海世博会非洲联合馆艺术总监。

Hu Xiangcheng
Famous Contemporary Artist. He has taught at the Shanghai Theater Academy and University of Tibet. He was one of the founders of Shanghai Biennale and worked for Shanghai Expo African Joint Pavilion as the Art Director.

主办单位 SPONSOR

上海市浦东新区人民政府
People's Government of Pudong District, Shanghai

承办单位 UNDERTAKER

上海浦东新区新场镇人民政府
上海浦东土地控股（集团）有限公司
上海新场古镇投资开发有限公司
Shanghai Pudong Xinchang Town People's Government
Shanghai Pudong Properties (Group) Co.,Ltd.
Shanghai Xinchang Town Investment Development Co.,Ltd.

协办单位 SUPPORTERS

上海市浦东新区规划和土地管理局
上海市浦东新区文化广播影视管理局
上海市浦东新区商务委员会
Shanghai Pudong District Planning and Land Administration Bureau
Shanghai Pudong District Culture, Radio, Film & TV Administration
Shanghai Pudong District Business Council

支持单位 SUPPORTERS

上海戏剧学院
上海交通大学
同济大学
Shanghai Theater Academy
Shanghai Jiao Tong University
Tongji University

地点 LOCATION

第一展馆："更·新场"[上海浦东新区新场镇新场大街301-305号]
第二展馆："拾久新场"[上海浦东新区新场镇东后老街6号]
Exhibition Hall 1: Regeration · New Place [Xinchang Street No.301-305, Shanghai Pudong Xinchang Town]
Exhibition Hall 2: Memories of Xinchang [Houlao Street No.6, East of Shanghai Pudong Xinchang Town]

新场古镇
Xinchang Town

生活美学,点亮百年街巷
愚园路历史文化风貌区实践案例展
Lightening Time-Honored Street with Life Aesthetics
Site Project of Yuyuan Road Historic Area

愚园路初辟于清宣统三年 (1911 年),系公共租界工部局越界辟筑之路,以东端当时沪上名园愚园而命名。东起常德路,西迄定西路,全长 2.7 公里,跨静安、长宁两区。沿线弄巷空间丰富,建筑类型多样,景观丰富多变,既有风格各异的花园住宅、新式里弄、别墅公寓等居住建筑,亦有中西合璧的建筑群。如著名的西园大厦 (愚园路 1396 号 "九层楼")、宏业花园、沪西别墅 (愚园路 1210 弄,亦称 "好莱坞弄堂")、王伯群旧居、联安坊、桃源坊、瑞兴坊、歧山村等。

虽历经半个多世纪的演化变迁,愚园路整体历史风貌依然清晰可见。本街区核心保护范围面积约为 117.5 公顷,占街区总面积的 52.7%。有上海市级文物保护单位 4 处、区级文物保护单位 1 处,共有优秀历史建筑 170 余幢。

这里曾驻留过一批著名学者、政治家、艺术家、企业家、军官将领及工商巨子等,这些名人不仅影响着曾经的历史岁月,也为世人留下了宝贵的文化财富。正是这些珍贵的近代历史建筑和丰富的人文内涵使得愚园路历史文化街区成为上海中心城内规模较大、优秀历史建筑数量较多的历史文化风貌区,集中体现了上海西区近代华人高级住宅区的居住生活和以教育建筑为代表的公共建筑群的风貌特征。

但在 20 世纪八九十年代 "破墙开店" 热潮下,愚园路沿线形成违章搭建、街道杂乱、绿化景观设施破旧、商铺 "小、乱、散" 等现状,仅能满足周边居民的日常生活需要,为改变现状、提升片区的整体形象,长宁区政府将 "愚园路" 改造作为城市更新之案例,对于愚园路区域历史建筑的保护和再利用保持相当的前瞻性和开放性，使其不仅仅停留于居住功能，还融合内在的历史底蕴，开发更多样化的建筑功能，引入艺术、创意、体验式复合型的生活美学业态，保护更新愚园路历史风貌区，经过 2 年多的筹划、近 1 年的施工，于 2015 年 9 月初步形成愚园路"历史、精致、文化、融合"的街区新形象。

故园旧梦新路奇缘

作为沪西高级住宅最多的地区之一，愚园路街区浑然天成的雅致与神秘尤为让人沉醉。诸多历史名人曾居住于此，如爱国人士沈钧儒、文坛三剑客施蛰存、戏剧表演艺术家祝希娟、中国航天事业的奠基人钱学森、著名钢琴演奏家顾圣婴、原锦江饭店创始人董竹君等。斑驳的墙壁雕刻下岁月的痕迹，回旋的楼梯透传着昔日的气息，显赫的身世与叠加的故事酝酿。树影婆娑的法国梧桐，曲曲折折的幽深弄堂，写满了过往与曾经的优雅辉煌。自由奔放的巴洛克风格建筑任性地表达着对世俗的爱慕，威严高贵的老公馆体现着北欧风情的点滴格调，清水红砖的花园洋房演绎着英伦乡村风情……传承了一脉优雅血统的愚园路，如今呈现出一番生活美学的艺文街区风采。

通过政府和企业一系列的改造计划，愚园路融合了艺术、创意、文化等元素，加入了体验式复合型业态，顺应万众创新的当今潮流，引进硅谷式的联合办公空间、优质创业企业孵化器、艺术家国际交流空间、生活创意品牌集合店、极具独立精神的专业买手店、精致餐饮等。

例如，愚园路 1282 号原本是一家美甲店和小日用品店，如今已完美蜕变为一家体量虽小却具有国际化视野的设计师品牌概念店——REGALO（意大利语意为"礼物"），汇聚来自意大利、德国、法国、瑞士等独具特色的服饰设计师的创意产品。店内另有德裔匠师 Julia 现场手工打造首饰的工作台，workshop 的形式提供全手工打造高品质创意婚戒的服务，唤醒大众对"定制生活"的认知。从配搭到设计，从思想到行动，REGALO 以独立、自由的姿态在愚园路绽放，持续传递着爱生活、爱艺术、匠心独运的设计理念。穿越世纪的变迁，摩登时代将在「ART 愚园」重现。游走，发现，体会，一同走进故园旧梦，探索新路奇缘……

Case Exhibition Place – the Past and Present

Yuyuan Road, built in 1911 (the third year of Emperor Xuantong's Reign of Qing Dynasty), was constructed by Municipal Council in the foreign settlement and named after the then famous Yu Garden at its east end. It stretches 2.7 km from Changde Road in the east and to

上图　西园公寓
ABOVE Xiyuan apartment

左图　长宁区少年宫
LEFT Children's Palace of Changning District

Dingxi Road in the west, crossing Jing'an District and Changning District. The road is lined on either side by alleys, buildings and landscape of all kinds and styles, including garden house, lane house, villa and apartments, as well as building blocks fusing Chinese and Western features. List of famous buildings: West Park Mansions (A nine-floor building at No. 1396, Yuyuan Road), Hongye Garden, Huxi Villa (Lane No. 1210 Yuyuan Road, also known as "Hollywood Lane"), Former Residence of Wang Boqun, Lian'an Lane, Taoyuan Lane, Ruixing Lane, Qishan Country, etc.

The overall historical scenes of Yuyuan Road are still vivid even after more than half a century of changes and evolutions. The core protection zone covers 117.5 hectare, about 52.7% of the total block area. Here is home to 4 Shanghai Municipality Cultural Heritage Sites, 1 District Cultural Heritage Site, and 170+ Outstanding Historic Buildings.

It used to accommodate a number of famous scholars, politicians, artists, entrepreneurs, officers and business giants, who played key roles in history and left precious cultural heritage to us. These valuable historic buildings of modern times and their rich cultural implications make Yuyuan Road Historic Area a concentration of outstanding historic buildings above scale. Here presents high-end residence life of modern Chinese and the features of public architectural complex represented by education buildings.

Following the "Removing walls for Storefront" trend in the 1980s and 1990s, Yuyuan Road was occupied by illegal constructions, shabby greening facilities, and messy scenes, and it could only meet dailyneeds of nearby residents. To promote the image of the district, Changning District Government tookthe "Yuyuan Road" Transformation Project as a case ofurban renewal. Buildings along the Yuyuan Road have been protected and re-used with a forward-looking and open mind. On one hand, their housing functions are guaranteed, and on the other hand, diverse architecture functions are developed with their historic heritage. Industry mix of art, creative, experience is introduced to protect and renew Yuyuan Road Historic Area. After planning over two years and construction of almost one year, Yuyuan Road unveiled its new image of "History, Culture, Fusion, Refinement" in September of 2015.

上图　建筑内部更新 愚园路 1282 号
　　　REGALO 珠宝服饰店

ABOVE Interior renovation No. 1282, Yuyuan Road Jewelry store of REGALO

上图　1282–1288 号原商铺
ABOVE Original shops at No.1282 to No.1288

A New Start

With the highest density of luxury houses in Western Shanghai, Yuyuan Road is of irresistible charm for its natural elegance and mystery. Here is home to many historical figures, like patriot ShenJunru, Shi Zhecun, one of The Three Musketeers of literary circle, dramatic artist Zhu Xijuan, QianXuesen, founder of China's space cause, GuShengying, famous pianist, and Dong Zhujun, founder of Jinjiang Hotel. Age has left its trace on the old wall, and the corkscrew staircase triggers memory of the past glory and stories. Exuberant platanus trees and deep lanes reflect the once elegance and golden times. The Baroque architectures embody our deep love for this world, and the magnificent old mansion takes us to North Europe, while the red-brick garden house brings British Country to China. Yuyuan Road, elegant and noble from its birth, now takes on another image of everyday life aesthetics.

Thanks to a series of transformation programs of the government and enterprises, factors like art, creative, and culture are combined and experimental industry mixture is introduced here. Following the current trend of "Innovation By All", here embraces co-working space like that of Silicon Valley, quality startup incubator, international art exchange space, original brand concentration, independent buyer shops, and fine catering.

For example, No. 1282 of Yuyuan Road used to be a manicure and general merchandise store. Today, it became a designer brand shop, REGALO (meaning gift in Italian). This store, small in size but global in vision, gathers creative products by fashion designers from Italy, Germany, France and Swiss. At here, customers can buy hand-made jewelry by German craftsman Julia on her workshop in the store. Quality, creative wedding rings made purely by hands awake the public's awareness for "customization". REGALO, independent and free, silently shines on Yuyuan Road for its excellence in fashion matching, design, thinking and action. It promotes the design idea of loving life, loving art, and inventive products. The modern times come back at ART Yuyuan. Let us walk, discover and experience this new start of Yuyuan Road…

上图、右页　音乐市集现场
ABOVE & FACING PAGE Musical Bazaar

展览主旨

作为"上海城市空间艺术季"活动计划的一部分,"愚园路历史文化风貌街区更新改造"实践案例展将直观表达城市建设、街区改造的成果,通过愚园路文献展,梳理愚园路人文历史,对比改造前后风貌;同时,开展"愚园路艺术音乐季",形态多样的音乐艺术嘉年华,让艺术走入城市空间,进入社区,以多样的形式演绎"城市空间艺术"这一主题,与大家分享城市品质提升、城市空间重塑的新型规划建设成果,结合城市更新为街区注入极具活力的艺术体验。

展览介绍

"社区医院"

当代艺术展——"社区医院"作为"ART 愚园"上海城市空间艺术季活动案例展之一,在位于愚园路 1086 号的艺术交流空间展出。参展的艺术家呈现探讨哲学内在精神、个人及社会问题的一系列作品。艺术能唤醒人们的意识,帮助宣泄情绪,甚至引发人们的共鸣。我们将原社区医院作为探索、体验、定义甚至治愈"问题"的地方。这些跨媒介的艺术作品各自反映不同的"问题",部分艺术作品将是首次呈现。

"愚园路文献展"

"愚园路文献展"梳理愚园路发展脉络,整理并集中呈现愚园路人文历史文献、影像资料、原貌与现状对比图片等,以展览形式重现愚园路传奇故事、人文历史、建筑风格等。该文献展在江苏社区第四网格居民区活动中心展出。

"西方早期玻璃艺术展"
于愚园路 1250 号举办的西方早期玻璃艺术展,以实践和观展的形式全方位展示玻璃艺术的学术及技术知识。

左图　快闪活动
LEFT　Flash mob

右图　社区医院开幕现场
RIGHT　Opening scene of Community hospital

活动

实践案例展开幕仪式暨草坪音乐会
为配合愚园路的整体功能与品质的提升,长宁区政府和愚园文化创意发展有限公司举办了系列活动,让艺术走进生活空间,结合国际化的形式,拓展居民的文化视野,逐步实现融合联动。愚园路空间艺术展开幕仪式邀请了来自拉丁美洲的咖·丽贝乐队,为大家带来爵士、蓝调以及中国音乐元素与拉美风情巧妙结合的表演。

"和平颂歌·草坪音乐会"作为"上海城市空间艺术季·愚园路历史风貌街区案例展"的重要活动之一,于 8 月 27 日在愚园路 1107 号弘基创邑国际园草坪举办。上海馨田交响乐团以一曲庄严的《义勇军进行曲》拉开了音乐会的序幕,怀旧感与艺术感在"ART 愚园"交叠融合。音乐会以交响乐、声乐、朗诵等多种艺术形式,并邀请江苏社区的抗战老兵来到现场,向他们致以了崇高的敬意,充分展现了"ART 愚园"高雅的文化气息。

"愚园路音乐艺术季"
愚园路街巷里弄的转角,安排不同艺人进行演奏和表演,充分调动愚园路空间资源,形成符合街区特性、空间错落有致的丰富表演。通过快闪形式让观者在愚园路惊喜遇见生活之美的动人触点。午后阳台上的一段悠扬萨克斯,或是黄昏街角突然响起的提琴独奏,重启愚园路的静谧过往,重塑愚园路的繁华今生。另有音乐主题市集活动,把来自群众、体现本真的艺术注入城市空间,关联艺术与城市空间,通过互动表演形式带动愚园路游客和居民的情绪,驱散沉闷和单调,传播与大众息息相关的生活之美。

上图　和平颂歌·草坪音乐会活动现场
ABOVE　"Paean of Peace" Lawn Concert

下图　《社区医院》当代艺术展
BELOW　"Community Hospital" Contemporary Art Show

愚园路改造节点

城市空间艺术的根本点在于它是一种集生活、人文、建筑、历史、设计于一体的多元化产物,面对各民族、各年龄段、各文化层次的普通大众,"愚园路历史风貌街区"将城市更新概念与历史建筑保护相契合,将可见的物理空间与无形的文化底蕴叠加,展现出多元的城市空间。"愚园路历史风貌街区"一大特点就是纵横交错的弄堂,弄堂尽头的贯通,也就将愚园路的一段段历史串连了起来;愚园路上的一幢幢老建筑,宁静而安详,它们都是历史的传承者和见证者,像打开话匣的老者,讲述着上海城市发展的篇篇故事。

愚园路历史风貌街区更新改造实践案例展现的不仅有老弄堂、老洋房的保护,引入的新兴业态也具有极大亮点,使整条街道焕发出全新的光彩。愚园路1107号创邑国际园、少年宫和工人俱乐部景观节点、"新联坊"景观节点改造完成,引入的商户也将开展形式各异的文化沙龙及艺术展示,结合社区、游客举办文化类演出活动(音乐周、街头艺人表演、创意市集、音乐剧)。到场嘉宾共同为此次愚园路空间艺术展的开幕注入了美好的源泉,这也是 ART 愚园以"上海城市空间艺术季"为平台的首次正式亮相。同时,为配合愚园路的整体功能与品质的提升,创邑投资·愚园文化举办了一系列活动,让艺术走进生活空间,以国际化的形式结合中国传统经典文化,拓展了居民的文化视野,实现了融合联动。

上图 社区医院海报
ABOVE Poster of Community hospital

下图 社区医院展览现场艺术家与观众交流
BELOW Communication between Artist and audience in Community hospital exhibition

Theme of Case Exhibition

As part of "Shanghai Urban Space Art Season", "Practice Case Exhibition of Yuyuan Road Historic Conservation Area" will show results of urban construction and community transformation. Yuyuan Road Literature Exhibition helps to present Yuyuan Road's culture and history in an organized manner by comparing site features before and after transformation. Meanwhile, Yuyuan Road Art & Music Season is held to let art into urban space and communities, and the "urban space art" theme is interpreted via various forms. We deliver to citizens the achievement by new planning and construction that improves urban quality and reshapes urban space. We take advantage of this urban renewal opportunity to bring dynamic art experience.

Exhibition Introduction

"Community Hospital"

Contemporary Art Show –"Community Hospital" held in Shanghai Art Exchange Space (No. 1086, Yuyuan Road) is one of "ART Yuyuan" exhibitions during Shanghai Urban Space Art Season. At the show, a series of works which discuss internal spirit of philosophy, personal problems and social issues will be on display.

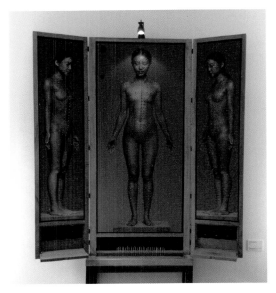
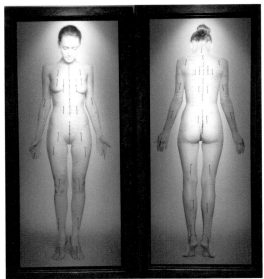

上图　艺术家常易的作品　　下图　愚园路文献展现场
ABOVE Works by artist Chambenoit Christian　　**BELOW** Literature Exhibition of Yuyuan Road

Art can awaken our dormant consciousness, get out our negative emotions and even strike a responsive chord in our hearts. For this reason, we bring these cross-media works which reflect different "issues" in this former community hospital, helping visitors to explore, experience, define and cure "problems". It is worth noting that some works will be premiered at this show.

Yuyuan Road Literature Show

Based on the history of Yuyuan Road, Yuyuan Road Literature Show presents legendary stories, cultural history and architectural styles of Yuyuan Road by collecting literature, videos and paired photographs before/after transformation concerning cultural history of Yuyuan Road. The show is held at the Residents Activity Center in the fourth grid of Jiangsu Community.

Early Western Glass Art Exhibition

The Early Western Glass Art Exhibition held at No. 1250, Yuyuan Road presents the information and technical expertise of glass art through practices and display.

Activities

Yuyuan Road Space Art Exhibition Opening Ceremony & Lawn Concert

To highlight the upgrade of Yuyuan Road in overall function and quality, Changning District Government and Shanghai Yuyuan Cultural Creativity Development Co., Ltd. launch a series of activities. With international cultural elements, these activities expand cultural vision of Shanghai citizens, and bring high art to common life, gradually achieving the synergy effect. For example, a famous Latin American band is invited to the opening ceremony of Yuyuan Road Space Art Exhibition to present Jazz, Blues and performances which ingeniously combine Chinese music with Latin elements.

上图　"西方早期玻璃艺术展"及现场活动
ABOVE "Western early glass art exhibition" and field activities

As one of important activities of Shanghai Urban Space Art Season · Practice Case Exhibition of Yuyuan Road Historic Conservation Area, the "Paean of Peace" Concert is held at the lawn of HongjiChuangyi International Park (No.1107 Yuyuan Road) on August 27. With solemn national anthem of China presented by Shanghai Xintian Symphony Orchestra, the concert kicks off in an environment mixed with nostalgia and artistic sense. At the concert, different artistic performances like symphony, vocal music and recitation are presented, and veterans of the Anti-Japanese War in Jiangsu Province are also invited and paid with great respect. In a word, the concert fully demonstrates strong and refined artistic ambience of ART Yuyuan.

Yuyuan Road Musical Art Season

By fully leveraging space resources of Yuyuan Road, different performers are arranged at corners of lanes and alleys along Yuyuan Road to present diversified shows which correspond to block features, for example, flash mobs which allow visitors to surprisingly experience exciting moments, and a piece of saxophone music at the balcony in the afternoon or a piece of violin solo at the street corner during the sunset which takes listeners back to the tranquil past of Yuyuan Road and gives a little surprise to busy Yuyuan today. Besides, a music-themed bazaar is held to incorporate mass art into urban space, and builds a connection between them. With interactive performances, the bazaar boosts emotions of visitors and residents, sweeps away tediousness and dullness, and promotes the beauty of life.

	1	
3	2	
	4	

1   少儿图书馆改造前
    Children's Library before transformation
2   少儿图书馆改造后效果图
    Children's Library after transformation
3   弘基创邑国际园改造前停车场
    Parking lot before transformation in Hongqi chuangyi International Garden
4   弘基创邑国际园改造后的公共草坪
    Public lawn after transformation in Hongqi chuangyi International Garden

上图　工人文化宫改造前
ABOVE Workers Cultural Palace before transformation

下图　工人文化宫改造后效果图
BELOW Workers Cultural Palace after transformation

Yuyuan Road Reconstruction Nodes

Urban space art, in essence, is the outcome of life, culture, architecture, history and design. Targeting at the general public of all ages and culture levels around the world, "Yuyuan Road Historic Area" combines urban regeneration with the protection of historic buildings and visible physicals pace with invisible culture to present diversified urban spaces. The criss-crossed network of lanes and alleys is a big feature of "Yuyuan Road Historic Area". With interconnection of these lanes, stories of Yuyuan Road are linked. Along Yuyuan Road, old buildings stand quietly and peacefully. As inheritors and witness of Shanghai history, they retell every chapter in Shanghai urban development.

Besides the protection of old lanes and foreign-style houses, "Yuyuan Road Historic Area" also introduces new exciting commercial activities, making the road regenerate with radiance. With the completion of reconstruction nodes like Chuangyi International Park (No.1107), Youth Center, Workers Club and "Xinlianfang", merchants will hold different cultural saloons and artistic exhibitions, together with cultural performances (music week, street performance, creative bazaar and musicals) focusing on communities and tourists. Besides, special guests are invited to help facilitate grand opening of this art space exhibition. This is also official debut of "ART Yuyuan" at the platform of "Shanghai Urban Space Art Season". At the same time, to highlight the upgrade of Yuyuan Road in overall function and quality, Chuangyi Investment · Yuyuan Culture launches a series of activities. By combining international cultural elements with Chinese classic culture, these activities expand cultural vision of Shanghai citizens, and bring high art to common life, achieving the synergy effect.

策展人感想

作为曾经的沪西高级住宅地区之一，愚园路浑然天成的雅致与神秘尤为令人沉醉。树影婆娑的法桐，曲曲折折的幽深弄堂，写满了过往的优雅辉煌。"上海城市空间艺术季"系列案例活动之一的"愚园路文献展"娓娓道来这深深弄堂的前世今生。"愚园路文献展"梳理愚园路发展脉络，整理并集中呈现愚园路人文历史文献、影像资料、原貌与现状对比图片等，以展览形式重现愚园路传奇故事、人文历史、建筑风格等，"愚园路文献展"将在江苏社区第四网格居民区活动中心长期展出。让我们穿越世纪的变迁，在"愚园路文献展"中一同体悟愚园路的前世之故，今生之味。

Note from the Curator

Yuyuan Road, one concentration of brownstone communities in West Shanghai, is of irresistible charm for its natural elegance and mystery. Exuberant platanus and deep lanes reflect the once elegance and golden times. "Yuyuan Road Literature Show", an activity of "Shanghai Urban Space Art Season" will tell you stories of these deep lanes. Based on the history of Yuyuan Road, the Show presents legendary stories, cultural history and architectural styles of Yuyuan Road by collecting literature, videos and pairedphotographs before/after transformation about cultural history of Yuyuan Road. Yuyuan Road Literature Show will be held at the Residents Activity Center in the fourth grid of Jiangsu Community. Let's leap through time to experience the past and present of Yuyuan Road.

策展执行团队 CURATION TEAM
孙嘉林 马骥　长宁区规划和土地管理局
黄志伟 赵光宇　上海愚园文化创意发展有限公司
Sun Jialin Ma Ji　Changning District Urban Planning & Land Authority
Huang Zhiwei Zhao Guangyu　Shanghai Yuyuan Cultural Creativity Development Co., Ltd.

主办单位 SPONSOR
长宁区人民政府
Changning District Government

承办单位 UNDERTAKER
长宁区规划和土地管理局
Changning District Urban Planning & Land Authority

协办单位 SUPPORTERS
长宁区文化局｜长宁区江苏路街道党工委办事处｜上海愚园文化创意发展有限公司
Changning District Culture Bureau/Party Committee Office of Jiangsu Road Changning District/ Shanghai Yuyuan Cultural Creativity Development Co., Ltd.

地点 LOCATION
长宁区愚园路（东至江苏路、西至定西路）
Yuyuan Road, Changning District (Jiangsu Road to the east and Dingxi Road to the west)

文中图片均由上海愚园文化创意发展有限公司提供
All photos are provided by Shanghai Yuyuan Cultural Creativity Development Co., Ltd.

愚园路上一家更新业态后的店铺
The shop after updating the commercial format in Yuyuan Road

对话
世博会城市最佳实践区实践案例展
Dialogue
Site Project of Expo Urban Best Practice Area in Huangpu District

世博城市最佳实践区原为工业厂房,于世博会期间对其进行了改造。世博会后,城市最佳实践区得到了完整保留,延续世博"美好城市"的主题,被打造成为集创意设计、交流展示、产品体验等为一体的,具有世博特征和上海特色的文化创意街区。

The Expo Urban Best Practice Area was originally an industrial plant, which was transformed for the Expo. It has been preserved after the Expo, continuing the theme of Expo "Beautiful City", so as to build a cultural and creative block with characteristics of the Expo and Shanghai, integrating creative design, exhibition and exchange and product experience.

展览主旨

本次实践案例展就城市更新问题以"对话"为主题,展开专家、学者、管理者、艺术家以及公众之间的国际对话、跨界对话以及公众对话等系列活动。以更新规划走进生活,走进社区为目的,在实践者与不同人群之间展开对话,实现东西互通、跨界互联以及社会共治的目标。

展览介绍

此案例展包含了"城市变奏曲"——上海城市设计联盟更新实践展、佩西城市更新实践成果展、中意手工艺匠精神巡展等三个展览。

城市变奏曲

以"城市变奏曲"为主题的上海城市设计联盟更新实践展,将上海城市设计联盟成员单位的城市更新实践案例作为主要内容。城市不同时期的变革与更新,犹如一曲变奏曲,同一城市空间,在不同的时代主题下进行着迥然不同的创新与演绎。展览期间邀请了广大市民共同参与讨论城市更新话题,搭建设计师与公众对话的平台。展区面积约400平方米,以展板和视频形式为主,共展出30个典型案例。城市不同时期的变革与更新,犹如一部变奏曲,同一城市空间,在不同的时代主题下进行着迥然不同的创新与演绎。为了展现城市的变奏主旋律,本次展览按照展陈内容,分为一楼大型互动城市音乐装置区和二楼城市更新案例区两大板块。其中二楼的城市更新案例区,又根据案例主题依次划分为老城区功能提升、工业区更新以及公共空间更新三个展区以此展示不同的城市乐章。

值得一提的是,本次展览空间所在的上海设计中心,其前身原本是浦江之畔的工业厂区,经过世博会最佳实践区改造和世博会后的功能更新,如今演变为兼具城市时尚和空间创意的设计策源地,这座建筑的本身就是上海城市更新的见证者。

老城区功能提升更新——上海城市最佳实践区
城市最佳实践区,是上海世博会的一大创新项目,世博期间集中展示全世界最具创新性的城市宜居实践。借世博会契机,城市最佳实践区进行了两次更新。一次更新,整合土地资源,建设世博场馆;二次更新,在世博会后,延续其基本格局,并对部分建筑进行改造和新建,传承"Better city, Better life"理念,形成文化创意街区,成为上海城市发展的有机组成部分。

实践区位于世博园区、黄浦江畔,占地面积15.08公顷,是呈现2010年上海世博会主题的重要载体之一。以同济大学唐子来教授领衔的设计

上图 "城市变奏曲"展览现场
ABOVE "Urban Variation" exhibition

对页 "城市变奏曲"展览平面
OPPOSITE Plan of "Urban Variation" exhibition

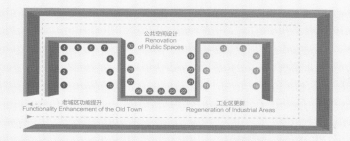

1 上海城市最佳实践区
——未来城市范本
The best urban practice area in Shanghai – future urban template

2 上海市黄浦区董家渡社区（C010 401 单元）控制性详细规划
Regulatory plan of Dongjiadu community (Unit C010401), Huangpu district, Shanghai

3 申都大厦改建工程
Reconstruction project of Shendu Mansion

4 利物浦一号
——利物浦中心城市更新项目
Liverpool No. 1 – renewal project of central city in Liverpool

5 创智天地
Knowledge and Innovation Community (KIC)

6 RTL 莱茵厅
RTL Rhein hall

7 上海"一城九镇"之北美风情金山枫泾规划
"A city of nine towns" – planning of Fengjing town

8 上海青浦城区中心区规划与实施
Planning and implementation of central Qingpu district, Shanghai

9 上海大厦
Shanghai Mansion

10 Vodafone 沃达丰办公总部
Vodafone headquarter

11 上海杨浦区黄浦江北岸工业旧址更新的概念城市设计
Renewal of old industrial site in north coast of Huangpu River, Yangpu district, Shanghai

12 上海闸北区市北高新园区整体转型及更新规划
Overall transformation and planning of Shibei high-tech zone, Zhabei district, Shanghai

13 上海杨浦滨江文化活力走廊城市设计
Urban design of cultural vitality corridor in Binjiang, Yangpu district, Shanghai

14 世博源、世博庆典广场及和兴仓库
Expo river mall, Expo celebration square and Hexing warehouse

15 EXPO 世博村
Expo village

16 株洲清水塘生态工业新城规划
Planning of ecological and industrial town in Qingshuitang, Zhuzhou

17 桃浦科技智慧城
Taopu smart city

19 陆家嘴人文绿网
Lujiazui cultural green grid

20 苏河湾地区景观风貌规划研究
Planning and study of landscape features in Suhewan

21 上海黄浦江沿岸——新华、民生、洋泾码头地区实施规划
Huangpu river coast – planning of Xinhua, Minsheng and Yangjing docks

22 上海后工业景观示范园
Demonstration park for Shanghai post-industrial landscape

23 上海滨江森林公园
Shanghai Binjiang Forest Park

24 社区街道 DIY——塘桥社区街角空间更新改造参与式规划
Community street DIY – renewal and reconstruction for street corners of Tangqiao community

25 轨交 6 号线站点地区空间重塑
Space rebuilding of stations in Line 6

26 青岛德国企业中心
Germany enterprise center in Qingdao

27 新城市画卷——新建路艺术改造
New city scroll – artistic reconstruction of Xinjian road

28 新诗意小镇——唐镇公共空间策划
New poetry town – planning of public space in Tangzhen town

29 国际诗歌小镇——李白与国际诗人的对话
International poetry town – dialogue between Li Bai and international poets

30 意大利米兰中央火车站
Central railway station in Milan, Italy

左图　"城市变奏曲"展览现场
LEFT　"Urban Variation" exhibition

团队，从多方面对城市最佳实践区进行了会间和会后两次更新规划与设计。

世博会期间，城市最佳实践区涵盖了能源利用、文化遗产保护、水资源保护、老城保护和发展、新农村建设、城市交通等一系列城市发展问题的解决案例，是世博会历史上的一个创举，并成为了城市建设管理和专业人士交流、分享、推广城市最佳实践的全球平台。

世博会后，城市最佳实践区后续发展以文化创意产业为主题，融合商务办公、文化艺术、会议展览、商业餐饮、休闲娱乐、酒店公寓、开放空间为一体。城市最佳实践区延续世博会期间的基本建筑格局，并对部分建筑进行改造和新建。

老城区功能提升更新——创智天地
项目地点：上海市杨浦区
设计单位：（美国）SOM建筑事务所

从2003年开始，瑞安房地产与上海政府合作，在杨浦区五角场打造创智天地。SOM建筑设计事务所借鉴了美国硅谷的成功经验，发展了集"工作、生活、学习、娱乐"于一体的综合性规划。经过12年的发展，原有的工业区已转型发展成为体现校区、社区和园区三区联动的知识型社区。由大型跨国企业、民营企业和初创企业构成的企业生态链业已形成。作为知识社区的社交中心——大学路连接了大学与园区，构建了富有活力的街道。历史建筑江湾体育场也被修缮并焕发出新的生机。

工业区更新——EXPO世博村
项目地点：上海市浦东新区
设计单位：（德国）HPP亨派建筑设计咨询有限公司

上图 "城市变奏曲"展览现场
ABOVE "Urban Variation" exhibition

下图 "佩西城市更新实践成果展"现场
BELOW Pesch Renewal Practice exhibition

世博村建于上海浦东东北部的规划区内，占地44公顷。HPP在国际设计竞赛中脱颖而出，获得了总体规划设计一等奖，并作为2010年世博会重大项目之一得以实施。

在黄浦江边的地块拥有得天独厚的区域优势，是现代化城市规划的重中之重，其手法是将原位于市中心的工业废区转化成为一个充满活力而又不失文雅的可持续发展城区。充分发掘了本地块的各种潜质后，利用多种功能叠加的设计方法在地块内设有酒店、公寓、购物和文化休闲等设施。

公共空间更新——陆家嘴人文绿网
项目地点：上海市浦东新区
设计单位：(美国) AECOM公司

陆家嘴人文绿网是针对上海浦东陆家嘴金融城全区提出的城市活力提升框架，以开放空间为载体，透过空间改造与完善，达到提升城市活力的整体效益，该框架整合公共交通、街区文化、公共活动节点、生态网络、城市地标、及绿地等功能网络。

透过上而下的空间叠图与下而上的人群观察，人文绿网从看似已无空间的都市丛林指认可改造的重点空间及提升策略，而后更聚焦小陆家嘴区，针对该区提出具体的空间设计导引，用以指导陆家嘴后续的空间设计与改造项目。

中意手工艺匠精神巡展
为促进中国工业转型，提升工业产品的审美品味，联合佛罗伦萨孔院举办以"传承，无界，干支，唤醒艺匠精神"为理念的巡展交流。此展览将使传统手工艺文化得以再现，以此唤起设计师、城市居民对文化沉淀和艺匠精神的重新思考。展区面积约200平方米，以艺术品实物的形式为主，共展出10余组艺术作品。

德国佩西作品展
联袂德国ppa|s（佩西）设计事务所举办佩西城市更新实践成果巡展，通过对德国代表性的城市更新规划实践、教学科研的深度展示，提供一个思考和讨论城市发展中如何应对变革、转型提升的平台。本展览已于11月27日开幕，邀请了斯图加特大学教授——佩西来到上海与公众展开对话，与中国同行和上海市民分享德国规划界在城市转型与城市更新领域的实践经验。展区面积约400平方米，以展板和模型的形式为主，共展出40余个更新实践案例。

本页　"意匠精神"手工艺展品
THIS PAGE　"Craftsman Spirits" Craft Exhibition

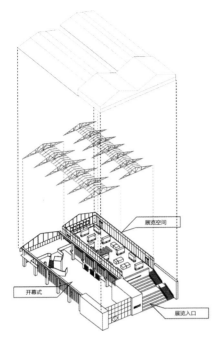

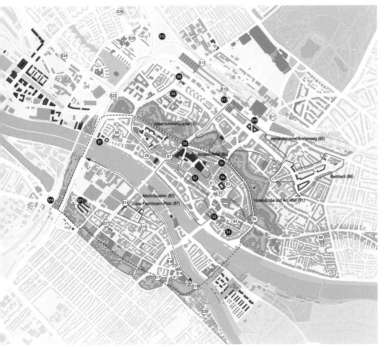

		4	
1	2	5	6
3		7	

1　展场布置图
　　Layout of the exhibition

2　"佩西城市更新实践成果展"展览现场
　　Pesch Renewal Practice exhibition

3　不来梅内城空间发展战略规划2025
　　City spatial development strategic planning of Bremen for 2025

4　科隆环城景观大道更新规划
　　Renewal planning of landscape ring avenue of Cologne

5　德累斯顿历史"新城"2025规划
　　"New Town" planning for 2025 of Dresden

6　不来梅内城鸟瞰
　　Aerial view of Bremen

7　德累斯顿历史"新城"鸟瞰
　　Aerial view of "New Town" of Dresden

74

左图　"佩西作品展"展场入口
LEFT　Entrance to Pesch Exhibition

论坛

世博实践案例展共举办了三场论坛活动，包括 SEAHI 论坛、"对话2015——中德城市更新论坛"和中意文化之旅论坛。

SEAHI 论坛

第二期 SEA-Hi! 论坛联袂 2015 上海城市空间艺术季举办。英国皇家特许建造学会中国区主席孙继伟、上海大学美术学院院长汪大伟、上海音乐学院教授王勇、IBM 亚太区智慧城市建设总规划师岳梅樱、原作设计工作室主持建筑师章明和"跑步猫体能训练"品牌创始人季斐翀等 6 位城市不同领域的嘉宾，从不同视角，就如何把人文艺术融入城市公共空间这一话题，畅谈自己独特和精彩的想法，共同挖掘这个城市潜在的无穷创意，跟大家分享空间艺术体验，畅想城市更新理念。孙继伟作了题为"用群心的光芒点亮城市"的演讲，他认为，经过 30 年的高速发展，对数量和速度的追求应该成为过去，而对品质的追求应该成为一种城市自觉。政府作为城市建设的主体，应当担负更多责任，但同时城市发展也需要开发商、设计师和市民的共同参与。

汪大伟从艺术家的角度，阐释了公共艺术对城市空间和居民的重要价值。他认为公共空间产生的问题有物理空间、行为空间和社会空间三个方面，地方精神的重塑是公共艺术的核心理念，因地制宜的重塑方式是公共艺术的方法论，影响力的重塑是公共艺术的重要评价指标。

岳梅樱在"智慧城市顶层规划方法"中为我们介绍了智慧城市的建设思路。她认为，要借助感知化、互联化与智能化改造现有城市环境，通过云计算、大数据、移动互联、社交网络四大核心技术的持续运用，营造有吸引力的产业环境和高质量的生活环境，带动整个经济发展框架的转型升级，提升市民幸福感，并有效缓解城市发展压力。

王勇则以"音乐让城市更美好"为题，为我们讲述了音乐地标与城市文化的关系，并指出世界音乐地标对于城市文化建设的重要性。他建议，上海也应该建设自己的音乐地标，并且应该向更为专业化的方向发展，同时要尊重音乐家，从而以更加智慧的方式保护上海的城市文化根脉。

季斐翀以亲身经历告诉我们，跑步已成为一种新的生活方式。他提议将修建塑胶跑道、设置跑步路线的配套设施作为发展体育产业的具体内容，同时作为中心城区旧城改造的重点内容，改善居民生活质量，提升城市竞争力。

章明以上海南市发电厂的前世今生为例，探讨了建筑再生的问题。他指出，老建筑留下了城市的记忆，因此我们要积极利用老建筑中的文化资本，尽可能保留其空间特征和历史信息，同时注入新的功能，以适应社会需求。

对话 2015——中德城市更新论坛
2015 年 11 月 27 日在上海设计中心南馆举办了"对话 2015——中德城市更新论坛"。来自同济大学的都市历史文化遗产保护专家张松教授主持了本次活动。德国的规划界权威、德国国家城市发展政策委员会委员、佩西城市发展咨询的创始人弗兰茨·佩西教授作了名为"城市设计的艺术"的主题演讲。演讲在理论阐述的基础上，通过展示"科隆环城大道"、"鲁尔区多特蒙德凤凰湖"等富有魅力的城市更新案例，向中国同行及社会公众分享了德国在城市更新领域的宝贵经验。随后，来自中国城市规划设计研究院的袁海琴、英国 BDP 姜齐冰、Lew James 与荷兰 KCAP 事务所的陈亚馨也分别以"中央活力区的更新规划思考——大城市与小城市的对比分析"、"迈向城市更新的框架与策略"和"伦敦沃克斯豪九榆树巴特西区域改造 —— 新考文特市场"为主题报告共享了其在中欧城市更新领域中的研究成果。

"中意文化之旅 —— 无界·传承"论坛
11 月 28 日，在上海设计中心南馆，由中意设计创新中心、同济大学中意学院、上海城市空间季联合举办了"艺匠精神"展之"中意文化之旅 —— 无界·传承"论坛。

著名市井人物画家谢友苏、新闻界高级摄影师兼纪录片传播人顾泉雄、意大利艺术歌剧家及美声理论研究者亚历山德罗·帕塔里尼受邀出席。上海百老德育讲师团成员、同济大学师生现场聆听了三位大师分享的艺术文化人生之旅。

上图 开幕式现场表演
ABOVE Performance at Opening Ceremony

下图 "城市变奏曲"展览
BELOW "Urban variation" exhibition

左图 "城市变奏曲"展览互动音乐装置
LEFT Interactive installation at the "Urban variation" exhibition

论坛上,同济大学中意学院副院长、中意文化交流项目负责人刘东教授介绍了同济对意合作及在文化建设方面取得的成果,指出中意文化交流、传承与唤醒文化思考的重要性。上海百老德育讲师团威泉木团长随即也发表讲话,指出艺术、城市、人与精神文化始终是分不开的,鼓励大家为艺术无界与文化传承做出贡献。

Theme of Case Exhibition

In this case exhibition, with the theme of "Urban Regeneration" by the way of conversation, we carried out a series of activities including international conversations and cross-border dialogues among experts, scholars, managers, artists and citizens. On the purpose of bringing the renewal plan into life and community, we carried out dialogues between practitioners and different groups of people, to achieve the objectives of mutual communication, cross-border exchange and social governance.

Exhibition Introduction

The exhibition includes: "Urban Variation": Shanghai Urban Design Alliance Renewal Practice Exhibition, Pesch Urban Renewal Practice Exhibition and Itinerant Exhibition of Sino-Italian Craftsman Spirits.

Urban Variation

The theme of Shanghai Urban Design Alliance Renewal Practice Exhibition was "Urban Variation". As the main content, the practical cases of urban regeneration provided by members of Shanghai Urban Design Alliance showed the urban renewal and regeneration in different periods, as if a piece of music was performed into a set of variations under the same urban space but different themes of the times with innovation and development. During the exhibition, the citizens would be invited to jointly participate in the discussion of urban renewal, to build the platform for dialogue between designers and the public. With about 400 m², the Exhibition was displayed in the form of panels and videos, and 30 typical cases were exhibited in total. Changes and evolution of a city during different ages are performed like variations;

the same urban space has been experiencing dramatically different innovation and evolvement under different age themes. To display the major melody, the exhibition covers two parts, large musical installations for interactive activities on the 1st floor and cases of urban regeneration on the 2nd floor. The latter is divided into three themed areas, namely, functionality enhancement of the old town, regeneration of industrial areas, and renovation of public spaces.

Shanghai Design Center, where this exhibition is located, used to be an industrial plant next to Huangpu River. After the transformation at World Expo to the best practice area and further functionality promotion ever since, the space has now become a source for designs characterizing both fashion and space creativity. The building itself witnesses Shanghai's regeneration.

Functionality promotion of the old town - the best practice area in Shanghai
The best practice area is an innovative project of Shanghai World Expo, where the most creative practices of city habitability in the world are displayed. And the area underwent two regenerations during and after the Expo, one for integration of land resources for pavilions, and the other for building and renovating some buildings for the concept of "Better city, Better life" to become blocks of cultural creativity and an organic part integrated into Shanghai's development.

Located in the Expo zone and next to Huangpu River, the practice area covers 15.08 hectares and served as one of the major themes at Shanghai Expo 2010. It was designed by a group led by Tang Zilai, professor of Tongji University, and re-planned and redesigned during and after the Expo.

During the Expo, lots of solutions to urban development in terms of energy consumption, cultural heritage preservation, water resource protection, protection and development of the old town, new rural construction and urban traffic could be found in the best practice area. It is a pioneering work in the Expo's history, which becomes a global platform for experts in city construction and management to communicate, share and promote the best practices.

After the Expo, with continuity of the existing architectural pattern and building and renovatoin of some buildings, the best practice area, under the theme of cultural creativity, has integrated business offices, culture and art, conference and exhibition, catering, entertainment, hotel apartment and open spaces.

Function promotion of old cities - Knowledge and Innovation Community (KIC)
Location: Yangpu district, Shanghai
Designer: SOM (U.S.)
Since the beginning of 2003, Shui On Land Limited, in cooperation with Shanghai government, has been engaged in the construction of KIC at Wujiaochang, Yangpu district. By imitating the successful concept of Silicon Valley, SOM has merged "work, living, study and entertainment" in the plan. 12 years' effort has made the original industrial plant into a knowledge community with colleges, neighborhoods and an industrial park. And the ecological business chain with multinational, private and newly-established firms has taken shape. The community's traffic hub, Daxue Road, connects the colleges and the industrial park, invigorating the whole blocks. And the historic Jiangwan Stadium has also been renovated with a modern concept.

Regeneration of industrial area - the Expo Village
Location: Pudong New Area, Shanghai
Designer: HPP (Germany)

The Expo Village, covering 44 hectares, is located at the planning area in northeast Pudong, Shanghai. HPP stood out and won the first place in the international design competition, and was appointed as the designer for one of the major projects in World Expo 2010.

The lot next to Huangpu River enjoys exceptional advantages, and is regarded as the most important place for urban modernization. And this can be fulfilled by turning the former downtown industrial wasteland into a vigorous and elegant urban district featuring sustainable development. Multi-functional design methods will be adopted for the facilities, including hotels, apartments, shopping malls, and cultural entertainments, after full exploration of the potential of this lot.

Renewals of public space - Lujiazui cultural green grid
Location: Pudong New Area, Shanghai
Designer: AECOM (U.S.)

Lujiazui cultural green grid focuses on the framework of urban vitality promotion proposed by financial center of Lujiazui; and in order to increase such vitality, the open public space is altered and improved. This framework integrates public traffic, block culture, public activities, ecological network, urban landmark, green lands and other functional networks.

Through top-down space overlapping and bottom-up observation of people, focus spaces are picked from the seemingly space-less urban jungle and improvement strategies will be made. Afterwards, the focus is back on Xiaolujiazui. According to the spatial design guide set up in such area, subsequent space design and project renovation can be performed in Lujiazui.

Itinerant Exhibition of Sino-Italian Craftman Spirits

To promote China's industrial transformation and enhance the aesthetic taste of industrial products, the tour exhibition, with the concept of "Heritage, Boundless, Stem & Branch and Artisan Spirit" was jointly held with Confucius Institute at the University of Florence. The reappearing of traditional handicraft culture aimed to arouse the designers and residents to rethink about the culture and artisan spirit. The Exhibition roughly covered 200 m^2, and displayed over 10 groups of works in the form of art objects. Changes and evolution of a city during different ages are performed like variations; the same urban space has been experiencing dramatically different innovation and evolvement under different age themes. To display the major melody, the exhibition covers two parts, large musical installations for interactive activities on the 1st floor and cases of urban regeneration on the 2nd floor. The latter is divided into three themed areas, namely, functionality enhancement of the old town, regeneration of industrial areas, and renovation of public spaces.

Shanghai Design Center, where this exhibition is located, used to be an industrial plant next to Huangpu River. After the transformation at World Expo to the best practice area and further functionality promotion ever since, the space has now become a source for designs characterizing both fashion and space creativity. The building itself witnesses Shanghai's regeneration.

Renewal Practice Achievement Exhibition of the German Master Pesch

The Renewal Practice Achievement Exhibition was jointly held with Germany Pesch & Partner, Architects and Urban Planners (p p a | s), and provided a platform of how to cope with urban renewal and transformation in the process of reflecting and discussing the urban development by deeply displaying Germany typical urban renewal planning practice and teaching and scientific research. The exhibition was inaugurated on November 27, and invited Pesch, the Professor of University of Stuttgart to have a conversation with the public in Shanghai. He shared the practice experience concerning urban transition and regeneration in German planning circles with Chinese designers and Shanghai citizens. Covering about 400 m², the Exhibition was displayed in the form of panels and models with above 40 renewal practice case in total. The duration is from November 27 to December 27 (Monday to Saturday).

Forums

There were three forums in the Expo practice case exhibition, including SEAHI Forum, Dialogue 2015: Sino-Germany Urban Renewal Forum and China-Italy Culture Forum.

SEAHI Forum

The 2nd SEA-Hi! The Forum was held in the Kapok Hall of Shanghai Design Center with Shanghai Urban Space Art Season 2015. Six distinguished guests expressed their unique views on how to combine the arts and humanities with urban public space from different angles and explored the urban potential creativity, as well as shared experience of space arts and urban renewal ideas with the public, including Sun Jiwei, Chairman of the Chartered Institute of Building (China), Wang Dawei, Dean of Fine Arts School of Shanghai University, Wang Yong, Professor of Shanghai Conservatory of Music, Yue Meiying, IBM Asian-Pacific Smart City Chief Planner, Zhang Ming, Chief Architect of Original Design Studio and Ji Feichong, Founder of Running Cat.

Sun Jiwei made a speech on "lightening the city with people's hearts"; in his eyes, people should pursue quality instead of quantity and speed in urban planning after China's 30 years of development. While developers, designers and citizens are necessary in urban development, the government, as the main actor of urban construction, should bear more responsibility.

Wang Dawei explained the importance of public art to urban space and residents with an artist's tone. In his view, public space could be divided into physical, action and social ones; reshaping the local spirit was the core concept of public art; methodology of the public art should vary according to local conditions; and reshaping the influence was the key evaluation index of the public art.

Yue Meiying introduced construction thoughts of smart cities in her speech on "Method of Top-level Planning of Smart Cities". She held that, the perceived, interconnected and intelligent concept should be used to remold the urban environment, and core technologies, such as cloud computing, bid data, mobile internet, and social networking, should be further used to create attractive industrial climate and high-quality living environment, and to bring transformation and upgrade for the overall development framework of economy, and to increase citizens' happiness, and to release the urban development tension.

Wang Yong delivered a speech on "music makes city better", in which he narrated the relation between musical landmark and urban culture, and pointed out the importance of the former to the latter. He suggested that a musical landmark should also be built in Shanghai, and it should a professional one which would mix the musicians' idea and be helpful to protect the urban culture in a more intelligent way.

Ji Feichong, according to his personal experience, told us that running had become a new life style. He proposed that plastic runways and supported facilities with set route should be developed in sports industry, and should be viewed as the key project for reconstruction of old cities, in order to improve the living standard and urban competition.

Zhang Ming discussed building regeneration through the past and future of Shanghai Nanshi Power Plant. He pointed out that old buildings were filled with urban memory; therefore, the cultural capital in them should be taken advantage of, their spatial feature and historic trace should be remained as much as possible, and new elements should be added into them to meet the social need.

Dialogue 2015: Sino – Germany Urban Renewal Forum

Dialogue 2015 - Sino-Germany Urban Regeneration Forum was held in South Pavilion of Shanghai Design Center in November 27, 2015. Zhang Song, professor in Tongji University and expert in urban historical and cultural heritage protection, presided over this event. Professor Franz Percy, authority of planning circle in Germany, member of National Urban Development Policy Committee of Germany and founder of PPAS Urban Development Consultant, made a speech on the art of urban design. He, based on theories, shared valuable experience of Germany in urban renewals among Chinese counterparts and social public, through the display of marvelous cases, such as "Kolner Ringstrae" and "Dortmund Phoenix Lake". And then, Yuan Haiqin from China Academy of Urban Planning and Design, Jiang Qibing and Lew James from BDP, and Chen Yaxin from KCAP, respectively shared their reports on research achievements in urban renewals in central Europe, including "Reflect of Renewal and Planning for Central Dynamic Area-Contrastive Analysis between Large and Small Cities", "Toward Urban Re-Generation: Framework and Strategies", and "Vauxhall Nine Elms Battersea Regeneration – New Covent Garden Market".

China-Italy Culture –" Boundless · Heritage" Forum

On November 28, the Forum themed as "Artisan Spirit" was jointly held by China-Italy Design and Innovation Center (CIDIC), Sino-Italian Campus of Tongji University and SUSAS in South Pavilion of Shanghai Design Center.

Some honored guests were invited to attend the Forum including Xie Yousu, the famous painter specialized in character painting, Gu Quanxiong, Senior Photographer and documentary spreader, as well as Alessandro Patarini, the Italian opera master and Bel Canto theoretical researcher. Members of Shanghai Hundred Old Moral Education Lecturer Group and teachers and students of Tongji University had the privilege of hearing the experience sharing on art and cultural life by the three masters.

On the Forum, Professor Liu Dong, Vice-president of Sino-Italian Campus of Tongji University and Principal of China-Italy Cross-culture Communication Project introduced the achievements on cultural development through the cooperation between Tongji University and Italy. He also pointed out the importance of cultural exchange, inheritance and awakening reflections on culture. Qi Quanmu, the Director of Shanghai Hundred Old Moral Education Lecturer Group gave a speech, stating that it was always inseparable among the art, city, as well as people and spiritual culture, and encouraged the public to contribute to the boundless art and cultural heritage.

策展人感想

石崧 上海市城市规划设计研究院

2015 年 12 月 15 日，为期三个月的 2015 上海城市空间艺术季在徐汇西岸主展馆正式落幕。此次盛会，"以艺术之名、行文化之事、汇设计之力、促空间之变"。由于很好地将市民诉求、城市趋势和政府导向完美地结合，当我年初向上海城市设计联盟的各个设计机构提出联合举办案例展参加空间艺术季的倡议时，得到了所有联络员朋友们的一致认同。在随后的几个月时间里，这种意识认同转化为实实在在的行动支持，从策展团队的组建、实践案例的提供，到布展方案的设计、办展经费的筹措，在这个分享经济的时代，所有参与人员一同亲身践行了一次众筹众创的行动，成功地在上海设计中心南馆举办了"对话——世博会城市最佳实践区实践案例展"。我想，能把方方面面的资源凝聚在一起举办这次盛事的秘密，恰恰在于"文化兴市，艺术建城"的艺术季理念，这是这座称之为魔都的伟大城市人文情怀的彰显，也将成为城市的一份永恒的荣光。

Note from the Curator

Shi Song (Shanghai Urban Planning and Design Research Institute)

2015 Shanghai Urban Space Art Season closed at West Bund Art Center in Xuhui District on Dec.15. The exhibition aimed at "performing cultural undertakings on behalf of art; promoting special changes through the combination of design". In early this year when I applied to design institutes in Shanghai Urban Design Alliance for such exhibition which would combine well the citizens' claims, urban development trend, and governmental orientation, my proposal was accepted unanimously. During the following months, their acceptance turned into practical efforts, including building team for event planners, providing cases, designing the exhibition program, and raising money for it. I personally took part in the whole process in times of economy sharing. Finally, an exhibition of "Dialogue-The Best Urban Practice Area in World Expo" was successfully held at the Center. In my opinion, its success in absorbing resources of various circles stemmed from the Fair's concept of "to prosper cities with culture and art". It manifested humanistic feelings in this magic city and it would be the eternal honor possessed by Shanghai.

"城市变奏曲"
——上海城市设计联盟更新实践展
Urban Variation – Shanghai Urban Design Alliance Renewal Practice Exhibition

策划团队 PLANNING TEAM
华东建筑设计总院跨界艺术中心｜上海市城市规划设计研究院
Cross-border Art Center of East China Architectural Design & Research Institute｜Shanghai Urban Planning and Design Research Institute

策展顾问 CURATING CONSULTANT
王辉 Wang Hui
INLINK 艺联中国公共文化艺术策略机构，创始合伙人，艺术总监，兼任上海当代艺术博物馆公共关系策略顾问
Founding Partner and Art Director of INLINK China's Public Cultural Art Strategy Company, Part-time Strategy Consultant for public relation of Power of Station of Art

支持机构 SUPPORTED BY
上海城市设计联盟
Shanghai Urban Design Alliance

佩西城市更新实践成果展
PESCH Urban Renewal Practice Exhibition

策展人 CURATOR
张宁 Zhang Ning
ppa|s 佩西城市发展咨询
ppa|s Urban Development Consulting

中意手工艺匠精神巡展
Tour Exhibition of Sino-Italian Craftman Spirits

策展人 CURATOR
郑嘉
LAB WORLD 联合创始人，创意总监，中意设计创新中心负责人
Zheng Jia, Co-founder of LAB WORLD, Chief Creative Officer and Principal of CIDIC.

卓敏
佛罗伦萨大学孔子学院负责人
Zhuo Min, the Principal of Confucius Institute at the University of Florence

"城市变奏曲"展览中的展品,从不同的角度可以看到不同的影像
Exhibits of the "Urban Variation" exhibition, showing different views from different perspectives.

小步伐，大改变
虹口区音乐谷实践案例展
Small Steps, Big Changes
Site Project of Shanghai Hongkou Music Valley

虹口区音乐谷文化创意产业集聚区位于虹口港、沙泾镇、俞泾浦三河交汇处，有石库门里弄社区和众多特色建筑，是具有历史风貌区、艺术创意区、传统居住社区特点的创意集聚区。

The Cultural Creative Industry Cluster Area of Hongkou Music Valley is located at the junction of Hongkou Creek, Shajing Port and Yujingpu River. There are many stone gates, lanes and alleys, as well as numerous heritage architectures, which is a creative cluster region featured as the historic style, art creativity and traditional residential community.

案例展展览主旨

本案例展创作的出发点和立足点是让艺术回归生活、回归社区,通过融入社区"更新与再生"的过程,探索特定的城市公共空间的艺术场所形式,并赋予个性化的场所精神。此次我们将上海市虹口区音乐谷作为激发公共空间活力的的特定样本和试点项目,在城市微型公共空间作出永久改变之前,用1:1的比例临时搭建的方式测试解决方案。同时,探索公开对话和社区参与,在城市改变过程中邀请现有和潜在使用者参与、直面他们的需求。

城市微空间复兴计划

Let's Talk 学术论坛的创始人和组织者,资深建筑学术杂志编辑戴春博士和知名建筑师俞挺博士联合发起了"城市微空间复兴计划",旨在发动具有社会使命感的设计师们"寻找并研究身边的失落空间"并"对身边失落空间进行设计",共同激活这些负面空间,让生存空间更积极,让社区空间更有活力,进而更好地协助政府和公共机构改善城市面貌,促进城市空间的健康发展。

与传统的甲方提出设计要求,设计师接受设计任务、进行设计项目的程序不同,该计划需要设计师自己去寻找、发现、研究,在这个过程中设计师很从容地思考一些问题。比如城市当中需要哪些活动可以让城市更有活力,哪些空间可以承载这些活动,最终人和空间之间建立情感联系,大家都爱这个空间,都需要它,这个空间才会被维护,才会有持久的生命力。城市微空间复兴计划将作为一个平台联合一批学者、专家对提交方案及研究成果进行审核评定,提出改进意见,并推动合适的方案政府立项或以民间共建等方式实现。

目前该计划已得到广大青年建筑师的积极响应,其中一些建造已经在 Let's Talk 团队和虹口区规土局共同推动下获得政府立项,进入建造阶段。其中,基于对音乐谷地区出入口研究的设计——"路亭"完成建造;基于对音乐谷区域水系与公共生活研究、对音乐谷区域与周边区域连接研究而完成的设计——"花桥"和"浮桥"正在推进。

应该说在艺术季期间,"议事"的 Let's Talk 正在走向"做事"的 Let's Work,真正成为推动城市空间更新的力量。

上图 海报
ABOVE Poster

左图	路亭合照
LEFT	Group photo

"路亭"案例（已建成）
- 引子：一棵大树和音乐谷的入口

路亭选址位于虹口区音乐谷地区西北角的四平路和新嘉路口，拥有一株据说是四平路上最高大的树，是绿化部门在四平路道路拓宽时特意保留下来的，街角还有一个小广场。但入口与外界的联系依旧被车水马龙的四平路切断，长期以来一直是音乐谷颇为隐蔽的一个入口。音乐谷地区管理公司一直期望通过一个既醒目又符合音乐谷背景的标志性入口导引装置。

- 现场调研：公共空间本身缺乏设计，为占用行为提供可能

设计师现场调研过程中发现，实际上该街角的城市公共空间被汽车维修装饰美容店占据，作为停车、洗车、清理和维修场地空间。通过网上的新闻调查，公众对此类行为有诸多不满，但是仅通过城市管理手段无法完全根治此类现象，反而变成了一种城市顽疾，严重影响了城市公共空间的生活品质，城管部门也普遍表示很无奈。

但不可否认的是，公共空间本身缺乏设计也为这种占用行为提供了可能性，让公共空间转向半公共空间，对公众而言也成为一个仅供过路的空间。作为项目主持设计师俞挺和童凌峰在现场就立刻有了第一个想法，是否可以通过入口装置的介入提高此类负面公共空间的品质？

- 解决对策：入口装置应作为公共空间利益的协调

入口装置在功能上不应回避城市公共空间被占用的既成事实，但需要谨慎选择设计切入点，避免各种利益矛盾的激化和爆发。所以，艺术装

组图　路亭
OPPOSITE & THIS PAGE Entrance Pavilion

置设计的第一步是从原有的花坛绿化改造入手，拓展原有的非机动车和步行道路空间，作为对被占用的公共空间的补偿。设计方案中还考虑满足其他功能，对音乐谷入口标志性的提升和空间导视作用，未来普通市民和周边居民对此类公共空间的自发使用等。

这个案例能够证明设计也可以作为一种协调手段，为未来其他地区的类似问题提供了一个新的思路。

• 设计策略："路亭"——一把温暖的刀，切开冷漠的周边环境

项目设计师——筑竹空间创始人童凌峰介绍了设计的初衷："我们希望路亭为周围的市民撑起一个歇脚和交流的场所，同时不是大张旗鼓地向洗车店要回属于市民的空间，这个过程我们认为是一个漫长的过程，而不是一个激发矛盾的过程。"

由现代木结构建造的动感十足的路亭也会弥合被花坛分开的人行道和街角的公共广场，设计师希望它以一个活泼和有生命力的形式唤起居民对场所精神的认识。设计师希望它以轻巧的姿态介入，避免原空间占用者的抵触，所以，设计师非常巧妙地预留了一个依然可以让车通过的大门，除了供过路市民和周边居民休息驻留，也给洗车店开了一个口子。现在从海伦路地铁站出来，你就可以远远望见"路亭"的醒目结构，它俨然已经成为一个进入音乐谷区域的引导标志物。

花桥案例（在建）
• 引子：花桥是一座轻盈的花园，邀你在水上驻足

过去的江南水乡，即使是小桥也很神气，它们趴在波光潋滟的水面上，弓着背，顶出一个个显眼的小山包，人们要先上去，再下来，于是才有"你站在桥上看风景，看风景的人在楼上看你"。远远看去，也是桥最分明，等人，约会，桥上见。

而现在的桥是给车走的，马路经过水面就是桥，桥不再是一个"地点"了，只需匆匆通过。我们想造这样一座桥，它首先是一个"地点"，而非"交通"——项目主持设计师亘建筑事务所创始合伙人范蓓蕾如是介绍"花桥"。

• 水系研究：有没有可能恢复一些水上的生活，重新建立人们与河浜的联系？

过去的江南，水网四通八达，交通运输，生产生活，都离不开大小河道；而那些较小的河道，称为河浜。由于人们生活的依赖，自然会生出主人的责任心，悉心维护。开埠后的近代城市，水路交通逐渐转变为陆路交通，填满河浜就

上图　花桥
ABOVE Flower Bridge

左图　驳岸研究
ABOVE Study on revetment

虹口港：沉寂
THE HONGKOU CREEK: SILENCE

河道和道路将这片区域切割成零碎的地块，加上复杂的社会状况，这里在如火如荼的城市建设中显得相对安静
Compared with surrounding areas, the creek-splited land parcels and complicated social-histotical condition was hardly welcomed by developers

成为扩展城市公路的主要手段。只有一些较宽的河道承担了公共运输的功能，更多的河道在建设中被填埋成公路，如陆家浜、肇嘉浜，最著名的是英法租界界河——洋泾浜，现在是延安东路。河浜和人的生活关系慢慢消失了，河浜的管理由个人生活需求也逐渐转化为公共事务。

虹口区南部的哈尔滨路附近，汇集了俞泾浦和沙泾港两条穿越虹口区的河道，水面宽15～20米，是典型的江南小河道的尺度。它的交通功能相对苏州河来说较弱，更像是传统的河浜，是容纳市民生活的地方，只是现在已经与城市生活隔离，高高的防汛墙让它成为一个自闭的存在。我们把桥选址在虹口港和沙泾浜交汇处，音乐谷老场坊这个区段。河道和道路将这片区域切割成零碎的地块，在如火如荼的城市建设中，这块地显得相对冷清。冷清的另一个原因是这里缺乏公共空间，防洪堤边上很窄，只能容下一个人行道，于是新建的步行桥就变成了一个契机，它们不仅仅是连接两岸的便利设施，也可以是城市生活中突然冒出的公共空间，成为有人气的地方，比如作为约会等人的地方。另一方面，这个选址有助于建立音乐谷地区与四川北路商圈透过武进路的连接，对该区域的活力激发有意义。

右页　"光之韵"景观灯光装置
OPPOSITE PAGE　"Light" landscape lighting installation

- 设计策略："柔软的花园轻轻地放在"防洪堤上

建筑师范蓓蕾对这个桥的姿态有个设定，它应该是个"柔软的花园轻轻地放在"防洪堤上，好像自己没有支撑，只是搭在堤上一样。步行桥中间好像由于自重，略微地弯下去了一点，从河岸上看，边缘非常薄，有点虚幻，就像是一个浅浅的托盘，盛着各种鲜花，还有上面的人。当然，实际上结构不可能只有边缘那么薄，最后建筑师把结构、形态和使用综合起来考虑，设计出了这样的效果。栏杆安装在花池的内侧，这样，从侧面看，人的活动就像处在花丛里；再者，人靠近栏杆的时候，隔着花，离水面还有一段距离，会更有安全感。在花桥两端，我们加上轻质的楼梯，人们在这里要先上桥再下桥，虽然是防洪堤的原因，但是这一上一下，跟过去的桥还挺像的，也是一个小小的舞台。

乐器与灯光装置

通过融入社区"更新与再生"的过程，探索特定的城市公共空间的艺术场所形式，并赋予个性化的场所精神。关注城市公共空间审美形式与人的关系，展示社区的历史文脉、现在与未来，通过公共艺术与社会的参与，让艺术家、社区居民、园区企业、政府机构和游客，以一种互动的方式，在音乐谷文化创意产业园和历史风貌街区之中塑造有意味的城市公共空间。展览以艺术品实物的形式为主，共展出2组艺术作品：《小乐队》与《光之韵》。

展览介绍

设计作品方案展

通过社区调研图片、案例展策展创意方案和公共艺术作品设计方案，展示对城市更新的思考，推动音乐谷文化生态社区的再生和城市空间造境的探索。展览面积约200平方米，以图文展示的形式为主，共展出3个策展方案和6组公共艺术作品设计方案。

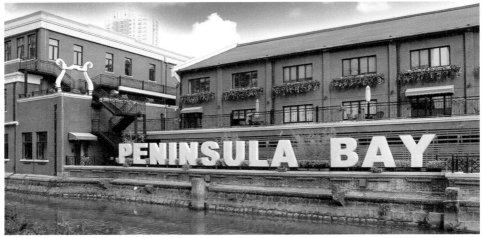

本页　"小乐队"铜管乐器装置
THIS PAGE　"Small Orchestras" Brass instrument device

左图 "城市微空间复兴计划"港大上海中心专场,约 200 人参与
LEFT Urban Micro Space Renewal Plan held in Shanghai Study Center of Hong Kong University with about 200 participants

城市微空间展

2015 年 12 月 21 日 – 2016 年 1 月 15 日期间,Let's Talk 系列学术论坛携手香港大学建筑学院上海学习中心、Wutopia 工作室、大舍建筑工作室、旭可建筑、GOA 大象设计、亘建筑事务所、筑竹空间和一批设计师在香港大学建筑学院上海学习中心一楼展厅举办"城市微空间展",回顾并总结一年的工作成果并对公众介绍"城市微空间复兴计划",引发更多的民众对城市空间更新的关注与参与。

论坛及活动

Let's Talk 论坛

作为沪上最活跃的非官方学术论坛,"Let's Talk"自 2014 年成立之初至今已成功举办 53 场活动。建筑、设计、艺术、互联网……名家、学者、求学者、职业人、爱好者……不同领域的嘉宾、听众因"Let's Talk"走到一起,通过讲座、对谈、互动,多元化的观点前所未有地在这里激烈碰撞。"Let's Talk"由沪上知名资深建筑人俞挺、戴春创立,最初由沪上一批知名建筑师开展讲座与研讨,逐步形成了创新思维涌现的话语平台;论坛以不定期邀请专家、学者嘉宾参与讲座、对谈等形式展开。

在 2015 上海城市空间艺术季期间,Let's Talk 学术论坛、虹口区规划与土地资源管理局共同主办以"城市更新"为主题的系列学术讨论,以虹口案例展音乐谷区域中的"旮旯空间"为主会场,香港大学建筑学院上海学习中心为分会场,组织了 13 场专题学术演讲,累计邀请各界专家、学者、艺术家近 70 位作为演讲与讨论嘉宾,场场爆满,引发各界讨论,成为社会热点。

左图	设计作品方案展现场
LEFT	Design Works Exhibition

右页　海报
RIGHT　Posters

9月—12月相关主题：
两种视角看城市——三位海归设计师的观点与实践
上海中心：重塑未来垂直城市
从北到南：哈建青年建筑师的探索与实践
建筑之路：探索E城E乡下的更新
张佳晶：聊宅志异
城市微空间复兴计划
说说我们建筑界的那些奖
天空之城——高密度建筑实践与图景
基础设施建筑学
抵达身体：现代性与反现代性的方式
城市如何长高——"上海高度"首讲
城市更新之罗昂对策——在既有城市环境中设计
更新与更新：上海城市更新计划
"城市微空间复兴计划"港大上海中心专场，约200人参与
"更新与更新：上海城市更新计划"嘉宾合影

文化生态社区的复兴研讨会
12月10日，中共虹口区委常委宣传部部长刘可、上海华夏文化创意研究中心理事长苏秉公、上海社会科学院历史研究所研究员郑祖安、上海石库门文化研究中心主任张雪敏、复旦大学社会学系教授于海、东华大学教授陈祖恩、教授级高级工程师俞挺、瑞安集团董事长助理周永平等50余位专家学者与社会各界人士，从城市更新的角度出发，深入剖析上海音乐谷等城市街区文化生态再生的成效和存在问题，研讨如何通过公共艺术等方式，在大规模城市改造之后修复尚存的城市街区文化传统风貌，在传承特有的历史文脉的同时，实现城市更新中的社区再生，

使不同的历史街区显现出不同的文化特质。

玩在"树与河"之间——音乐谷半岛湾创客艺术街集市
10月17 – 18日,音乐谷半岛湾创客集市在沙泾港河畔举行,来自沪上6所学校、9个院系的大学生带着艺术作品聚集于此,活动方沿着苏州河道设置了超过66个展位,展出1000多件高校艺术作品及社会创意人士作品。

本次活动的参展单位有同济大学设计创意学院、艺术与传媒学院、建筑与城市规划学院、华东师范大学美术系、上海大学美术学院、东华大学服装与艺术设计学院、上海理工大学版院支点工作室、上海工程技术大学艺术设计学院、中法埃菲时装设计师学院。活动不仅为青年艺术家提供创意作品的展示平台,推动艺术创客市场化,更为苏州河沿岸的文化创新提供了新航向,推动音乐谷在城市更新中进行社区再生。

Theme of Case Exhibition

This exhibition aims at bringing art back to earth and to communities; with art blending into the "renewal and regeneration" of the communities, artistic locations will be created with specific and personalized public spaces. We will take Hongkou Music Valley as the specific sample and pilot project to vitalize public space, and adopt a makeshift way for the urban micro public space before its permanent change, so as to test new solutions with a proportion of 1:1. Meanwhile, we will explore the open dialogues and community involvement, and invite the existing and potential users to participate in the process of city changes, as well as collect the ideas according to their own needs.

上图 俞挺博士在 Let's Talk 论坛上发言
ABOVE Dr. Yu Ting speaks at the Let's Talk Forum

下图 上海虹口区音乐谷眘旲空间 Let's Talk 讲座现场
BELOW Lecture of Let's Talk at the GALLERY Space in the Hongkou Music Valley area, Shanghai

Urban Micro Space Renewal Plan

With the development of a series of academic forums of Let's Talk, there are many topics related to the regeneration and renewal of public space among the lectures. As the founder and organizer of the forum, Doctor Dai Chun, a senior editor of academic architectural magazine and Doctor Yu Ting, a notable architect jointly launched the "Urban Micro Space Renewal Plan", aiming to motivate designers who have strong social responsibility, to discover, to research and to design the "lost space around them", so as to make the lost space full of vigor and our living space more positive, as well as better assist the government and public institutions to improve the look of our cities and promote the healthy development of urban space.

Differing from the traditional plan, that is, designers have to design projects as required by Party A, but this plan needs the designers to find, discover and research by themselves, and leave room for them to think about some problems. For example, what activities can make the city more energetic? What space can bear those activities? Eventually, it is likely to establish the emotional connection between people and space. Only when we love and rely on the space will it be maintained and survive for a long time. The Micro Space Renewal Plan

上图　参与讨论的艺术家们
ABOVE Artists in discussion

will serve as a platform to unite a group of scholars and experts to review and assess the submitted plans and research results, and then help push those appropriate project approved by governments or gain private assistance.

Currently, most young architects have positively responded to the plan, where some architectures have been under construction supported by the government with the joint efforts of Let's Talk team and Hongkou District Planning and Land Management Bureau, among which the design regarding the access to the Music Valley area – "Entrance Pavilion" has been completed; the "Flower Bridge" has been approved by the government based on the research design of the water system and public life in the Music Valley area, and the "Floating Bridge" has been under construction, which depends on design of connection in or around the Music Valley region.

In other words, Let's Talk is being turned into Let's Work during the art season, which really becomes the power to push the regeneration of urban space.

"Entrance Pavilion" Case (Completed)

Intro: A big tree and entrance to the Music Valley

The Entrance Pavilion is located at the junction of Siping Road and Xinjia Road in the northwest corner of the Hongkou District Music Valley area. There is a big tree there, which is said to be the tallest in the Siping Road. It is specially reserved by the Greening Department during the extension of Siping Road. Additionally, there is a small square in the corner. But the entrance to the outside is still cut off by the crowed Siping Road, so it has been a covert entrance to the Music Valley for a long time. The management company for the Music Valley area always expects to make it visible through an entrance guide device, which is also in line with the background of the Music Valley.

Site Investigation: Public space may be occupied for lack of design

By the site investigation, designers found that the public space in the corner is occupied by vehicle maintenance shops as car parking, washing, and cleaning and maintenance space. It was learned that the public made complaints about the actions according to online news. However, such phenomenon cannot be completely eliminated only by means of urban management, but evolve into a headache for the city, seriously impacting the life quality in urban public space. For this, the Urban Management Department generally expressed they were very helpless as well.

But it is undeniable that lack of design of public space also makes it possible for such occupancy. As a consequence, the space becomes a semi-public space, just as a passway for the public. On the scene, a new idea instantly came to the minds of Yu Ting and Tong Lingfeng, the designers of the Project. Whether can we improve the quality of such negative public space via entrance devices?

Solutions: The entrance device should be used as the medium for the interests of public space

It is an admitted fact that the urban public space has been occupied, but the design entry point needs to be carefully chosen for the entrance device, so as to avoid or reduce the conflicts of interests. Thus, we shall firstly transform the original flower bed and green space, and expand the space for the non-motor vehicle and pedestrian road, as the compensation for the occupied public space. Other functions have also been considered in the design plan, such as highlighting the entrance signs of the Music Valley, space guides, and spontaneous use by the coming ordinary citizens and residents around the public space, etc.

It is possible to prove the design is also available as a means of coordinated balance via this example, to resolve the interest conflicts between the government management department, park management department for the Music Valley, both of which represent the interests of the public, and the residential property or shop owners and the operator of car maintenance shop, representing the private interests. Additionally, it also provides a new idea for the similar problem in other regions in the coming days.

Design Strategy: "Entrance Pavilion" - a warm knife to make a way for the public

Project Designer – design concept introduced by Tong Lingfeng, the founder of Bambuspace: "We hope that the Entrance Pavilion can provide a place for nearby residents to stop for a rest or communicate with others, rather than aggressively claim the public space back from the car washing shop. We think it is a long process, not that to make conflicts."

The dynamic Pavilion made of modern wooden structure will also bridge the sideway separated by the flower bed and the public square at the street corner. Designers expect it can arouse the public awareness of the place spirit in a lively and vigorous way, and avoid the conflicts with the original occupant via a low profile. Therefore, they skillfully set aside a door for cars. It provides not only the rest area for passerby and residents, but an opening for car washing shop. Now you can see the eye-catching structure "Entrance Pavilion" if you come out from Hailun Road Metro Station, which has become a guiding sign into the Music Valley area.

Flower Bridge Case (Under construction)

Intro: Flower Bridge, a hanging garden invites you to enjoy the view on the water.

In the past waterside of south Yangtze River, even a bridge was willing to lie on the glittering water surface just like arching its back or supporting a tiny hill. If you went across the bridge, you would really experience the feeling of "To take in the view, on the bridge stood the viewer". Looked at from afar, the bridge was the most eye-catching. On it, you might wait for someone, meet or date with friends.

But now the bridge is crowed by vehicles, which is beside the road. People did not enjoy scenery on this "place", but rush through it. We want to build such a bridge as a location rather than a traffic facility. – Fan Beilei, the designer of the project and the founding partner of genarchitects, introduced the "Flower Bridge".

Water system research: is it possible to recover some life on the water? Or rebuild the relations between people and creeks?

In the south of the Yangtze River, the network of rivers expands in all directions, and people's life depend heavily on various river channels, including water transportation, production and

life; and the small river channels are so-called creeks. The dependence on rivers naturally leads to a sense of responsibility for local people to take care of them. However, after the opening of ports, the waterway transportation is gradually transformed into road traffic in the modern city, so filling creeks become the main means to broaden the urban highway. Only some wide rivers are used as public transportation, and numerous creeks have been filled during road construction, such as Lujia Creek and Zhaojia Creek. The most famous is the boundary river of British and French Concession – Yangjing Creek, which is the Yanan East Road now. The relations between creeks and human life are slowly disappearing, and the creek management is gradually transformed from personal life demands into public affairs.

Near to the Harbin Road in the south of Hongkou District, there were two river channels, Yu Jing River and Shajing Creek across the Hongkou District, with 15 - 20 m wide, which was the typical size of creek road in the south Yangtze River. Its traffic capability was weaker than that of Suzhou Creek, so it was more like a traditional creek, suitable to human living. But now it becomes a closed area, isolated by a high flood wall from the urban life. We selected the junction of Hongkou Creek and Shajing Creek, the zone of the old workshop in the Music Valley area as the location of bridge. This region was split into different land parcels, and the creek-split land parcels were hardly welcomed by developers. On the one hand, there is lack of public space. The flood bank is narrow, only available for a person to pass. Thus, the new footbridge becomes popular because it is not only the convenience channel to connect both sides, but the additional public space. It will be a place worth it, for example, dating or waiting for someone. On the other hand, the location will improve the connectivity between the Music Valley and commercial district of North Sichuan Road through Wujin Road, which is significant to stimulate the vitality in the area.

Design Strategy: A "soft garden" is "gently" placed "over" the flood embankment

Fan Beilei, the architect has special idea regarding the posture of the bridge. It should be a "soft garden" "gently" placed "over" the flood embankment. Seen from a far distance, it just seems to rest on the embankment without any support. The bridge will slightly bent down in the middle place due to its own weight. Seem from the bank, the edges are very thin and a bit unreal, just like a shallow tray full of flowers, as well as passers-by. Actually, the structure can be never as thin as the edge, and it is merely the design effect by integrating its structure, form and function. The handrails are installed in the flowers, so it seems for people to act in the cluster of flowers from the side view. Moreover, the bridge will be safer, for flowers are between people and handrails, and there is a certain distance from the water surface. At both ends of Flower Bridge, we add light stairs due to the flood embankment, and people shall go upstairs and downstairs, which is kind of bridge style in the past, like a small stage.

Instrument and lighting devices

With art blending into the "renewal and regeneration" of communities, artistic places will be created with specific and personalized public space. Focus on relation between aesthetic forms of urban public spaces and people; display the historical context, the present and the future; through the participation of public artists and the society, draw from interactions among artists, residents, enterprises, organizations and tourists, to create meaningful urban public spaces in the creative industry park of the music valley and historic blocks. This exhibition mainly focuses on real artworks, and two pieces of works, A Small Band, Rhythm of Light, are displayed.

左页　设计展空间
OPPOSITE　Design Exhibition

Exhibition Introduction

Design Works Exhibition

The pictures of community research, creative solution to case exhibition and design scheme of public art works are used to show the reflection on urban renewal, so as to promote the exploration to the regeneration of cultural eco-community in the Music Valley and artistic conception of urban space. Covering about 200 m², the exhibition will mainly be presented in the form of photos and illustrations with 3 curatorial plans and 6 groups of design plans of public art works in total. The duration is from November 15 to December 31.

Urban Micro Space Exhibition

From December 21, 2015 to January 15, 2016, designers of Let's Talk series of academic forums would jointly hold the "Urban Micro Space Exhibition" with those in Shanghai Study Center, the Faculty of Architecture, HKU, Wutopia Lab, Atelier Deshaus, Atelier Xük, GOA, Genarchitects and Bambuspace, as well as a group of designers on the first floor of Shanghai Study Center, the Faculty of Architecture, HKU. They also reviewed and summarize the one-year work results, and introduced the "Urban Micro Space Renewal Plan" to the public, so that more people are willing to focus on and take part in the urban space renewal.

Forums & Events

Let's Talk

As the most active non-official academic forum in Shanghai, Let's Talk has successfully held 53 events since its establishment in 2014. Experts, scholars, learners, professionals and amateurs fond of architecture, design, art and Internet, etc., as well as guests at different levels and audiences come together for Let's Talk, where various unprecedented viewpoints are mutually impacted and sparks of thought are inspired through as series of lectures, dialogues and interaction. Let's Talk was jointly founded by Yu Ting and Dai Chun, the renowned senior architects in Shanghai. To begin with, some lectures and seminars were held by a number of famous architects, and then a discourse platform for innovative thinking gradually came into being; in the forum, experts and scholars may be invited to join the seminar, conversations or other forms.

During the 2015 Shanghai Urban Space Art Season, the Let's Talk forum and Hongkou District Planning and Land Management Bureau have jointly organized a series of academic discussions themed "Urban Renewal"; with the "Gallery Space" in the Hongkou Music Valley area as the main venue and the Shanghai Study Center, Faculty of Architecture of HKU as the affiliated venue, 13 thematic lectures have been held. Roughly 70 lectures, scholars and artists from all walks of life were invited as the honored guests. Those lectures were so popular that many audiences were attracted, which has become a social hotspot.

Related topics from September to December:
On Cities from Two Perspectives – Views and Practices of Three Overseas Designers
Shanghai Tower: Reshape a Future Vertical City
From North to South: Exploration and Practice of Young Architects
Road of Architecture: Research on Regeneration of E Village in E City

Zhang Jiajing: House Chatting
Urban Micro Space Renewal Plan
Awards for Architecture
City in the Sky – Architecture – To High Density
Infrastructure Architecture
Approaching the Human Body: the Modern and the Anti-modern
How to Grow Taller – the First Speech – "Shanghai Height"
Urban Regeneration – Logon's Strategy – Design in Context
Regeneration and Renewal: Shanghai Renewal Plan

Seminar on Renewal of Cultural Ecological Community

On December 10, nearly 50 experts and scholars deeply analyzed the effect and existing problems on regeneration of cultural ecology in urban blocks such as Shanghai Music Valley from the perspective of urban renewal, and discussed how to renovate their traditional style after the large-scale urban renewal by the way of public art, including Liu Ke, the Minister of the Publicity Department of the CPC Hongkou District Party Committee, Su Binggong, the Director of Shanghai Huaxia Cultural & Creative Research Center, Zheng Zu'an, the Researcher of the Institute of History, Shanghai Academy of Social Sciences, Zhang Xuemin, the Director of Shanghai Shikumen Culture Research Center, Yu Hai, Professor of Department of Sociology, Fudan University, Chen Zuen, Professor of Donghua University, Yu Ting, Professor of Engineering and Zhou Yongping, Assistant Chairman of Shui On Group, as well as people from all walks of life, so as to realize the community regeneration in urban renewal and make different historical streets show different cultural traits in the process of inheriting unique historical cultures.

Enjoy yourself between the "trees and rivers" – Peninsula Bay in the Music Valley

On October 17–18, the Art Fair of Peninsula Bay was held along the Shajing Creek. The students from 6 universities and 9 colleges and departments in Shanghai gathered here with their art works. The organizer set more than 66 booths along the Suzhou Creek, exhibiting over 1,000 art works from undergraduates and social designers.

The exhibitors include College of Design and Innovation, College of Art & Media and College of Architecture and Urban Planning, Tongji University, Fine Arts Department, East China Normal University, School of Fine Arts, Shanghai University, Fashion Art Design Institute of Donghua University, Fulcrum Studio of School of Publication & Printing and Art Design, University of Shanghai for Science and Technology, College of Art and Design, Shanghai University of Engineering Science and International Fashion Academy. The Event not only provided a platform to display young artists' creative works and promoted the marketization for art designers, but offered a new direction for the culture innovation along the Suzhou Creek and pushed the community regeneration during the urban renewal in the Music Valley.

策展人感想

俞挺 戴春

2015上海城市空间艺术季期间，我们参与了虹口案例展的策展，在市规土局、虹口规土局和文创办等单位的支持下，我们不仅组织了"Let's talk"艺术季期间13场有关城市空间发展的专题讨论，而且启动了"城市微空间复兴计划"，其中第一个空间激活项目——位于虹口音乐谷地区的"路亭"建成；同时，"从Let's talk到Let's work：城市微空间复兴计划"展开幕，展出一系列正在进行的激活负面空间小、微空间更新案例。这项计划目前已得到广大青年建筑师的积极响应。我们深感参与艺术季活动仅仅是开始，城市小微空间更新的讨论和实践需要持续地推进，小步伐、大改变，大家共同推动的都市更新就在我们身边。

Note from the Curator

Yu Ting, Dai Chun

During the 2015 SUSAS, we participated in the curation of cases exhibition in Hongkou District. With the support of units including Shanghai Municipal Planning and Land & Resources Administration Bureau, Hongkou District Planning and Land Management Bureau and Shanghai Cultural & Creativity Office, etc. we not only organized 13 seminars on city space development during the Art Season of "Let's Talk", also launched the "Urban Micro Space Renewal Plan", where the first space renewal project was completed – the "Entrance Pavilion" located in Hongkou Music Valley area; at the same time, the Exhibition of "From Let's Talk to Let's Work: Urban Micro Space Renewal Plan" was inaugurated, displaying a series of ongoing renewal examples concerning negative space and micro space. The Plan has been positively responded by most young architects. We know it just the beginning of the Art Season; continuous promotion is needed for the practice and discussion on the renewal of urban micro space, so we should jointly contribute to the urban renewal step by step.

策展人 CURATOR

俞挺 Yu Ting
教授级高工，博士，Utopia Lab 主持建筑师、旮旯空间联合创始人。曾任上海现代建筑设计集团现代都市院创作所所长，总建筑师。
Senior Engineer, Doctor, Founder of Utopia Lab and Co-founder the Gallery Space. And he served as the Director and Chief Architect in Shanghai Architectural Design & Research Institute Co., Ltd. - Urban Architectural Design Institute.

戴春 Dai Chun
博士，《时代建筑》杂志运营总监和责任编辑、T+A 图书出版工作室主任、旮旯空间联合创始人。
Doctor, Managing Editor/Operating Director of Time + Architecture Journal, Director of T+A Publishing & Media Studio and Co-founder of the Gallery Space.

卜冰 Bu Bing
硕士，集合设计主持建筑师、旮旯空间联合创始人。
Master, Chief Architect of One Design Inc. and Co-founder of the Gallery Space.

毛毛 Maomao
艺术家，设计师和国际艺术策展人，世界女艺术家理事会上海分会副会长，上海 99 国际女艺术家中心艺术总监，上海交界艺术机构联合创始人。
Artist, Designers and International Art Curator, Vice-president of the World Female Artists Association (Shanghai), Art Director of Shanghai 99 International Female Artists Center and Co-founder of Shanghai Jiaojie Art Agency.

张雪敏 Zhang Xuemin
上海石库门文化研究中心主任、同济大学历史文化名城研究中心研究员。
Director of Shanghai Shikumen Culture Research Center and Researcher of Research Center of Historical and Cultural City, Tongji University.

主办单位 SPONSOR
虹口区人民政府
People's Government of Hongkou District

承办单位 UNDERTAKER
虹口区规划和土地管理局
Hongkou District Planning and Land Administration Bureau

协办单位 SUPPORTER
虹口区文化局
Hongkou Bureau of Culture

地点 LOCATION
上海音乐谷（四平路、海伦路、溧阳路、周家嘴路、梧州路）
Shanghai Music Valley

市民坐在空间艺术装置《路亭》中
People are sitting in the space art device "Entrance Pavilion"

行走·跨越
上海天桥专题实践案例展
Walk · Cross
Shanghai Footbridge Thematic Exhibition

天桥，上海城市建设发展的一个缩影。20 世纪 80 年代天桥建设的高峰期至今，上海城市人口快速膨胀，交通量剧增，新增土地资源日益稀缺，逐渐步入了城市更新阶段。市政道路建设逐步从关注汽车、着重道路建设，向关注步行、注重公共交通建设转变。天桥也经历了从建到拆到再建的过程，这正是上海城市功能不断提升、设施不断完善、建设水平不断提高的见证。

大统路老旱桥	1955 年建成，是上海最早的、唯一的、跨越上海铁路南北的人行天桥。拆除于 1987 年，同年上海新客站建设配套修建了天目西路至永兴路的大统路非机动车立交。
共和新路新旱桥	1957 年建成，曾是上海第一座跨越铁路的车行立交桥。1995 年因建南北高架路而拆除。
延安东路外滩天桥	1982 年建成，平面呈 U 形。1993 年，外滩交通综合改造工程实施时，将天桥跨越中山东一路、中山东二路的两段拆除，仅保留跨越延安东路一段。1997 年延安东路高架东端下行匝道建造"亚洲第一弯"后，跨延安东路人行天桥还在，10 年后外滩交通再行改造，在拆除"亚洲第一弯"时也拆掉了天桥。
静安寺华山路人行过街地道	1983 年建成，是全市建造最早的过街人行地道。2 号线开通后与 2 号线连通。

四平路大连路天桥	1983年建成，平面呈"口"字形。修建地铁8号线时改成地下车行隧道。
徐家汇天桥	1984年建成，平面呈弧形，跨越了华山路、衡山路、肇嘉浜路和漕溪北路等路口，连接了第六百货、食品店、艺术书店、花鸟商店、徐汇剧场、公交车站等各类设施。1988年，兴建地铁一号线时拆除。
东大名路海门路天桥	1984年建成，按不规则地形条件呈不等边扇形，2004年拆除。
武宁路东新路天桥	1984建成，天桥的三个顶端连接着呈"品"字形排列的武宁百货商店、第三食品商店和沪西工人文化宫。
徐家汇天钥桥路天桥	1985年建成，2007年拆除重建，随着商业业态、周边环境等的改变，现在将进行新一轮改造，形成连接周边商圈的二层平台。
南京路西藏中路天桥	1985年建成，天桥四个转角处各有两座转式步行梯伸向沿街的第一百货商店、荣华楼酒家、新世界商场及上海精品商厦内。随着南京东路步行街、人民广场地铁站枢纽建设需要，天桥被拆除，新建了一条贯通人民广场地铁枢纽站与南京东路、南京西路的地下景观人行道。
延安东路西藏路天桥	1985年建成，平面呈高脚酒杯形。1996年建设延安高架路东段时拆除，1998年重建。
共和新路中华新路天桥	1985年建成，呈Y形，1994年建设南北高架路时拆除。
南京西路石门路天桥	1986年建成，呈S形，2001年建设地铁2号线时拆除。
中山东二路金陵东路天桥	1993年建成，为配合外滩地区交通综合改造工程，于2008年拆除。

上图	行走·跨越——上海天桥专题案例展厅入口
ABOVE	Entrance of "Walk·Cross"

四川北路海宁路天桥	1999年建成，为配合四川北路商业与景观改造，于2008年拆除。
吴淞路海宁路天桥	1999年建成，现在仍在使用中。
河南北路海宁路天桥	1999年建成，现在仍在使用中。
天目西路恒丰路天桥	1999年建成，现在仍在使用中。
河南南路复兴东路天桥	2009年建成，是国内首座X形钢结构吊索人行天桥，现在仍在使用中。桥下车道上没有设置任何支撑物，为驾车者和行人提供了较宽阔的视野。
陆家嘴二层步行连廊	陆家嘴中心区二层步行连廊（一期）工程由"明珠环""东方浮庭""世纪天桥""世纪连廊"四部分共同组成。陆家嘴天桥不仅是人车分流的优秀解决方案，更打造出一处极具标志性的城市景观，建成后的陆家嘴步行连廊，俨然已成为一处城市观光圣地。

天桥,承载着一代又一代市民的城市记忆。岁月更迭,行走在天桥上的上海人的生活方式和需求逐渐改变,购物、游憩、餐饮等休闲活动在市民出行的比例中越来越高,出行的方式也发生了巨大的变化,城市生活越来越丰富,街道活力慢慢复苏,天桥上的故事和风景也愈来愈精彩,刻在了每一个人的回忆里。

未来,天桥将是城市公共空间的一部分,是多元功能融合的桥梁,街区活力链接的平台,立体步行网络的组成部分,城市公共视觉中的一道风景线。在活动高强度、建设高强度的公共活动集聚区,天桥不仅将承担行走的基本功能,也将串联起商场、办公楼、绿地、广场、公交站点等各种设施,串联起多样的人群、多元的活动,成为一个立体、开放、友好、艺术的魅力公共空间。

行走,跨越,共同的经历;参与,共享,永远的记忆。

Overpass construction in the 1980s up to now, Shanghai has ushered in the fast population expansion and the fast increase of the traffic volume, but it faces the increasingly scarce support of the new land resources, so Shanghai has gradually entered into the phase for the urban regeneration. Shanghai has to pay the gradual attention to the car passage in the municipal road construction, particularly with emphasis laid upon the road construction, the walk and the transformation of the public transport construction. Overpasses have witnessed the process from the dismantling to the reconstruction and also witnessed the continuous improvement of the urban functions, facilities and the construction level in Shanghai City.

Datong Road Old Overpass	It was built in 1955 as the earliest and only pedestrian overpass across the southern side and northern side of Shanghai Railway, but it was dismantled in 1987. In the same year, Datong Road Non-motor vehicle interchange from West Tianmu Road to Yongxing Road was constructed as the supporting facility to Shanghai New Passenger Railway Station.
Gonghe New Road New Overpass	It was built in 1957 as the earliest vehicle overpass across the railway in Shanghai but it was dismantled for the construction of South-North Viaduct in 1995.
Yan'an Road Bund Overpass	It was constructed in 1982, taking the "U" shape. Two sections of the overpass across Zhongshan Dongyi Road and Zhongshan Dong'er Road were dismantled for implementing the Bund Integrated Transport Reconstruction Engineering Project, but only the section across East Yan'an Road remained. After the construction of "Asia No. 1 Bend" in the down ramp at the east end of the viaduct bridge in East

	Yan'an Road in 1997, the pedestrian overpass remained in East Yan'an Road. But after 10 years, the Bund Traffic facilities were renovated and the overpass was dismantled while "Asia No. 1 Bend" was dismantled.
Pedestrian Underpass next to Jing' An Temple, Huashan Road	It was built in 1983 as the earliest pedestrian underpass in Shanghai. It was connected with Metro Line 2 after Metro Line 2 was opened to traffic.
Overall between Siping Road and Dalian Road	It was constructed in 1983, taking the "mouth-shaped" pattern and it was rebuilt into the vehicle underpass while Metro Line 8 was built.
Xujiahui Overpass	It was built in 1984, taking the arc-plane shape across Huashan Road, Hengshan Road, Zhaojiabang Road and Caoxi North Road and other crossings, connecting No. 6 Department Store, Food Store, Art Book Store, Flower and Bird Store, Xuhui Theater, bus stops and all other facilities. It was dismantled in 1988 when Metro Line 1 was built.
Overpass between East Daming Road and Haimen Road	It was built in 1984, taking the equilateral fan-shaped pattern due to the irregular topographic condition, but it was dismantled in 2004.
Overall between Wuning Road and Dongxin Road	It was built in 1984, the three roofs of the overpass were connected with Wuning Department Store, No. 3 Food Store and Huxi Workers' Palace, all of which took the top-and twin-side shape.
Xujiahui Tianyaoqiao Road Overpass	It was built in 1985 and dismantled for reconstruction in 2007. With the business development and the change of the surrounding environment, a new round of renovation will be conducted to form a 2-story platform connecting the surrounding business circle.
Overpass between Nanjing Road and Tibet Central Road	It was constructed in 1985. There were two rotary staircases in the four corners of the overpass, extending themselves into No. 1 Department Store, Ronghua restaurant, New World Mall, Shanghai Boutique Mall. In view of the demand for the construction of East Nanjing Road pedestrian Street and Renmin Road Plaza Metro Station Hub, the overpass was dismantled for the reconstruction of an underground landscape pavement running across Renmin Road Plaza Metro Station Hub, East Nanjing Road and West Nanjing Road.
Overpass between East Yan'an Road and Tibet Road	It was built in 1985, taking the flat stemmed goblet shape. It was dismantled in 1996 for the construction of east session of Yan'an Road Flyover but it was rebuilt in 1998.

Overpass between East Yan'an Road and Tibet Road	It was built in 1985, taking the flat stemmed goblet shape. It was dismantled in 1996 for the construction of east session of Yan'an Road Flyover but it was rebuilt in 1998.
Overpass between Gonghe New Road and Zhonghua New Road	It was built in 1985, taking the "Y" shape but was dismantled in 1994 for the construction of South-North Flyover.
Overpass between West Nanjing Road and Shimen Road	It was built in 1986, taking the "S" shape but was dismantled in 2001 for the construction of Metro Line 2.
Overpass between Zhongshan Dong'er Road and East Jinlu Road	It was constructed in 1993 but was dismantled in 2008 to support the Bund Area Traffic integrated renovation engineering construction
Overpass between North Sichuan Road and Haining Road	It was built in 1999 and was dismantled in 2008 to make room for Business and landscape renovation project in North Sichuan Road.
Overpass between Wusong Road and Haining Road	It was built in 1999 and is still in use.

上图　百年·跨越——徐家汇城市更新案例展展场

ABOVE The exhibition venue of 100 Years · Cross - Xujiahui Urban Regeneration Case Exhibition

Overpass between North Henan Road and Haining Road	It was built in 1999 and is still in use.
Overpass between West Tianmu Road and Hengfeng Road	It was built in 1999 and is still in use.
Overpass between South Henan Road and East Fuxing Road	It was built in 2009 as the first X-shaped steel structure sling pedestrian flyover in China and remains in use now. There are no upholders on the vehicle passageways under the flyover to provide the drivers and pedestrians with the broader vision.
Lujiazui 2-story Walk Corridor	Engineering Project of Lujiazui Central Area 2-story Walk Corridor (Phase I) is composed of "Mingzhu Ring", "Orient Platform Bridge", "Century Flyover" and "Century Corridor". Lujiazui Overpass is not only an excellent solution for the separation of people and vehicles, but also creates a landmark urban landscape. The completed Lujiazui Walk Corridor remains a tourist attraction in the city.

Overpass record the urban memories of the citizens from generation to generation. With the elapse of the time, the Shanghai people walking across the overpasses have gradually changed their lifestyle and demands as they more frequently travels across the overpasses for shopping, leisure and catering activities. Of course, their traveling models have changed greatly as the urban life becomes richer and richer. The streets have the stronger vitality and attraction, so the overpass stories and scenes are more and more brilliant and alluring, fully implanted into the memory of each person.

In the near future, the overpass will be an integral part of the urban public space, the bridge with diversified and integrated functions, the platform connecting with the street vitality as well as an integral part of the 3- dimensional pedestrian network and the great landscape in the urban public vision. In the public activity clustering area with more activities and higher degree of construction, overpasses not only have the basic pedestrian functions but also are connected with all facilities in shops, office buildings, green land, plazas, bus stops, etc., and are closely connected with various groups people for diversified activities to form a 3-dimensional, open, friendly, arctic and charming public space.

Walk, lead-forward and joint witnesses; participation, mutual sharing and unforgettable memory.

展览主旨

市政设施建设，是现代城市建设的重要组成部分。本次"行走·跨越"——上海天桥专题案例展以上海天桥为市政设施建设的着眼点，通过公众互动、图文展示、高科技多媒体介入等多元方式将城市更新中天桥几十年来的发展历史，时代赋予天桥的不同内涵，以及人们对未来天桥的展望等通过艺术化手段呈现给参观者，使市民更加了解市政设施建设，拉近市民与市政设施的距离，展现上海城市市政建设的魅力和水准。同时，通过案例展的平台，提高市民对天桥建设的关注度，广泛听取政府、企业、社区各层面的意见和建议，为创新规划管理机制进行实践。

展览介绍

"行走·跨越"——上海天桥专题案例展于 2015 年 09 月 29 日在西岸艺术中心左前方的独立展馆开展。领导、专家、主策展学委会专家也亲临展厅，对天桥展的构思、内容、展览形式给予了现场指导。

为响应 2015 上海城市空间艺术季的"城市更新"主题，本实践案例展就"城市更新"问题以"对话"为主题，开展了专家、学者、设计团队、管理者与市民之间的对话系列活动，共同探讨了"提升城市空间品质"等内容。

市政设施景观化案例

原先裸露在这块地上的是一个泵房和两个箱式变电站，但它与后方艺术氛围浓厚、现代感十足的西岸艺术中心相比，显得不是那么协调，于是我们的策展人在短短的 3 个月时间内，发挥聪明才智，结合实际地形，设计出了这个简约但不简单的建筑。所以这个场地本身就是市政设施艺术化、景观化改造的一个成功案例。

图文模型展示部分

以图片文字和模型展示的方式，从不同角度表达了天桥在不同时期的不同内涵。按照过去、现在、未来的时间轴线，从上海的第一座天桥，大统路老旱桥，到上海真正意义上的第一座人行天桥——延安东路外滩天桥，再到重视和谐统一、成为都市景观的河南南路复兴东路吊索桥，最后到串联商旅文、沟通你我他的徐家汇空中连廊，详细介绍了上海发展史上的近 30 座天桥。天桥是上海城市建设发展的一个缩影，承载着一代又一代市民的城市记忆，未来也将成为串联起各种人群、活动的公共空间。

《上海天桥》《行走》《箱变》三个纪录片

《上海天桥》：邀请 SMG 制作上海天桥发展变迁的纪实片《上海天桥》，于 10 月 31 号"世界城市日"在上海纪实频道中午 11 点首播，同时在"行走·跨越"——上海天桥案例展厅也正式播出。

上图　天桥展厅二楼展示厅
ABOVE Second floor exhibition hall of Bridge exhibition

下图　《上海天桥》纪录片片段
BELOW Documentary footage of *Overpass of Shanghai*

《行走》：市民对天桥建设发展看法的采访小短片，在"行走·跨越"——上海天桥案例展厅播出。

《蜕变》：展馆从无到有整个施工建造过程的纪录片，在"行走·跨越"——上海天桥案例展厅播出。

徐家汇分展点

为了使本次活动更加丰富立体，2015年12月04日至2015年12月06日，在徐家汇公园也开展了以城市更新为主题的徐家汇分展。在主展点展示内容的基础上，再次演绎了天桥发展与城市建设的紧密联系。

左图	百年·跨越——徐家汇城市更新案例展展板
LEFT	100 Years · Cross – Xujiahui Urban Regeneration Case Exhibition Boards

活动

"行走·跨越"——上海天桥专题案例展共举办了两场论坛活动。主题分别为"对话·天桥"和"徐家汇城市更新"。

"对话·天桥"座谈会
时间：2015年9月14日下午
地点：上海设计中心北馆

与会人员：同济大学教授、博导，城市更新与设计学科团队主持人庄宇、上海营邑城市规划设计股份有限公司副总经理曹晖、市规土局市政处副处长王磊、市政管理局原副局长黄兴安、市政总院总工马骉、城建院原总工黄锦源、城建院原总工陈炳生、城建院原总工李坚、市政规划院原总工沈雷达、市政规划院设计研究所所长祝长康、上海市城市规划设计研究院原总工苏功洲、日本设计株式协会上海代表处首席代表葛海瑛、Callison建筑事务所（中国）设计总监周婕、Gensler建筑设计事务所设计总监Shamim Ahmadzadegan、奥雅纳工程顾问（上海）有限公司总建筑师、副董事Leonard Milford、奥雅纳工程顾问（上海）有限公司高级规划师李鹏。

内容：座谈会以天桥作为上海城市更新中市政建设的切入点，以轻松的对话形式请20世纪80年代以来上海天桥的新老优秀设计团队坐在一起，分享上海天桥的发展历史以及对天桥建设的认识和展望。

上图	"对话·天桥"座谈会
ABOVE	"Dialogue-Overpasses" Seminar
下图	"徐家汇城市更新"座谈会现场画布记录
BELOW	Canvas record of "Xujiahui urban renewal" forum

"徐家汇城市更新"座谈会

时间：2015年12月4日 9:00–10:30
地点：徐家汇商城集团1号会议室

与会人员：市规土局市政处、市规土局公众参与处、上海城市公共空间设计促进中心、徐家汇街道、区规土局、旅游局、文化局、交警、徐家汇商城集团、商家代表、徐家汇社区市民代表。

内容：在徐家汇城市更新的主题背景下，由街道、区规土局、文化局、旅游局、街道、交警等从不同角度出发，共同探讨徐家汇未来的发展定位。同时商城集团、商家、市民代表等结合自身实际，提出对徐家汇地区发展的诉求和建议。

上图 天桥展厅内部展示墙和模型
ABOVE Display wall and models within the overpass exhibition hall

下图 天桥展厅内部照片展示墙
BELOW Photo display wall within the overpass exhibition hall

Theme of Case Exhibition

Municipal infrastructure construction is an integral part of the modern city construction. This "Walk · Cross" – Special practical case exhibition of overpasses in Shanghai will focus on the municipal infrastructure construction of the overpasses in Shanghai to endow the overpasses with the different connotations in the process of the urban regeneration and the urban development history in the past decades through the diversified means, particularly the public interaction, photo and text display and the introduction of the Hi-tech multimedia and the expectation of the people for the future overpasses through the art approaches so that the citizen can have more understanding of the municipal infrastructure construction to shorten the distance between the citizens and the municipal facilities to show the charm and high level of the municipal infrastructure construction in Shanghai. In the meanwhile, the case display platform will attract the attention of the citizens to the overpass construction and we can collect the comments and proposals from the government, enterprises and people from all walks of life and conduct more practice in creating the planning management mechanism.

Exhibition introduction

"Walk · Cross – Special practical case exhibition of overpasses in Shanghai" was conducted in the separate exhibition hall in front of West Bank Art Center on September 29, 2015. Leaders, experts and other experts from major exhibition sponsor committee arrived at the exhibition hall, giving the site guidance related to the conceptual design, contents and exhibition form of the overpass exhibition. This exhibition was opened to the citizens free of charge from September 30 to December 15, 2015.

In response to the theme of "urban regeneration" in 2015 Shanghai Urban Space Art Season, this practical case exhibition focused on the theme of "urban regeneration" issues and "dialog" to conduct a series of dialogues among experts, scholars, design teams, managers and citizens to jointly explore for the contents of "improving the urban space quality", etc.

Cases of municipal facilities landscape

In the past there was a pump room two box-type substations on the land plot, but it stood brightly off against the profound art atmosphere and the modern West Bank Art Center in its rear, seemingly inharmonious, so Curator brought its talents into full play and design the simplistic but not simple building in combination with the actual topography within a short period of three months. So this area is actually a successful case of the artistic and landscape-based renovation of the municipal facilities.

Part of the graphic-text and model exhibition

The graphic-text and model exhibition reflects the different connotations of the overpasses in various periods from different perspectives. From the past, current and future time-line, we can see the representation and harmonious unity of the overpasses from the first overpass in Shanghai, Datong Road Old Overpass to the first pedestrian overpass of the real significance

1	2
3	4
5	

1 天目西路恒丰路天桥
Overpass between West Tianmu Road and Hengfeng Road

2 延安东路西藏路天桥
Overpass between East Yanan road and Tibet Road

3 南京路西藏中路天桥
Overpass between Nanjing road and Tibet Road

4 四川北路海宁路天桥
Overpass between North Sichuan Road and Haining Road

5 南京西路石门路天桥
Overpass between West Nanjing road and Shimen Road

in Shanghai, the sling overpass between South Henan Road and East Fuxing Road and Xujiahui Space Corridor connecting the commercial premises for better communication with you and me. So it gives us a detailed introduction of nearly 30 overpasses in the development history of Shanghai. Overpass is the epitome of urban construction and development in Shanghai and records the urban memories of the citizens from generation to generation. In the future they will be connected with different groups of people as the public space for activities.

Three Documentaries, respectively "Shanghai Overpasses", "Walk in Shanghai" and "Xiangbian"

"Shanghai Overpasses", a documentary reflecting the changes and development of Shanghai overpasses, was officially premiered in Shanghai Documentary Channel at 11:00 on October 31 -World Cities Day, and was simultaneously shown in "Walk- Leap-forward"- Shanghai Overpass Case Exhibition Hall.

"Walk in Shanghai" is a short interview documentary reflecting the points of views of the citizens about the overpass construction and development, which was officially shown in "Walk- Leap-forward"- Shanghai Overpass Case Exhibition Hall.

"Xiangbian" is a documentary reflecting the entire engineering construction process of the exhibition hall from the scratch and was shown in "Walk- Leap-forward"- Shanghai Overpass Case Exhibition Hall.

Xujiahui Exhibition Area

The urban regeneration theme Xujiahui sub-exhibition was conducted in Xujiahui Park during December 4 to December 6, 2015 with a view to ensuring richer and more all-dimensional display based on the display contents of the main exhibition venue to represent the closer interrelation between the overpass development and the urban development.

Events

Two forums were conducted for "Leap-forward-Xujiahui Urban Regeneration Case Exhibition", focusing on the themes, respectively "Dialogue-Overpasses" and "Xujiahui urban regeneration" Peditatiunt voloristiur aut endit, ulpa ditem il ipsam fuga. Et doluptatius, cum dolor asinulparum voluptate voles consequiae et enit, velent, sitatur, eratur molorep elignati beat.

"Dialogue-Overpasses" Seminar

Time: Afternoon of September 14, 2015
Place: North Hall of Shanghai Design Center

Participants: Zhuang Yu (Professor and PhD tutor, chief leader of Urban Regeneration and Design Discipline Team of Tongji University); Cao Hui (Deputy General Manager of Shanghai Yingyi Urban Planning and Design Co., Ltd.); Wang Lei (Vice Director of Municipal Office, Shanghai Municipal Bureau of Planning and Land Resources); Huang Xing'an (former Vice Director of Municipal Management Bureau); Ma Nie (Chief Engineer of General Municipal Design Institute); Huang Jinyuan (former Chief Engineer of Urban Planning and Construction Institute); Chen Bingsheng (former

Chief Engineer of Urban Planning and Construction Institute); Li Jian (former Chief Engineer of Urban Planning and Construction Institute); Shen Leida (former Chief Engineer of Municipal Planning and Design Institute); Zhu Changkang (Director of Municipal Planning and Research Institute); Su Gongzhou (former Chief Engineer of Shanghai Urban Planning and Design Institute); Ge Haiying (chief representative in Shanghai Representative office of Nippon Design Center, Inc.); Zhou Jie (design director of Callison Architect Office <China>); Shamim Ahmadzadegan (Chief Director of Ggensler Architect Office); Leonard Milford (Chief architect and associate director of ARUP Engineering Consulting <Shanghai> Co., Ltd.); and Li Peng (Senior planner of ARUP Engineering Consulting <Shanghai> Co., Ltd).

右上　徐家汇空中连廊鸟瞰效果
ABOVE Aerial view of Xujiahui air corridor

右下　徐家汇空中连廊人视图
BELOW People view of Xujiahui air corridor

Contents: The seminar focused on the urban construction of overpasses in the urban regeneration in Shanghai, so the old and new excellent design teams were kindly invited for a dialog concerning the overpasses built in the 1980s to mutually share their understanding and expectation of the historic development and the overpass construction in Shanghai.

"Xujiahui Urban Regeneration" Seminar

Time: 9:00–10:30 on December 4, 2015
Place: No. 1 Meeting Room of Xujiahui Business City Group

Participants: Municipal Office and Public Participation Office of Shanghai Municipal Bureau of Planning and Land Resources, Shanghai City Public Space Design Promotion Center, Xujiahui Sub-street Office, District Planning and Land Management Bureau, Tourism Bureau, Cultural Affairs Bureau, Traffic Police Bureau, Xujiahui Business City Group, representatives of commercial entities and citizen representatives of Xujiahui Community.

Contents: Under the theme background of "Xujiahui Urban Regeneration", representatives from the sub-street office, district Planning and Land Management Bureau, Cultural Affairs Bureau, Tourism Bureau and Traffic Police Bureau, etc., jointly discussed the future development positioning of Xujiahui from the different points of view. In the meanwhile, Business City Group, business entities and citizen representatives, etc., put forward their appeals and suggestions in view of Xujiahui district development in combination with their actual situations.

策展人感想

市政工程建设是城市空间品质的重要保障，天桥是其中一个很小的元素，但同样对空间环境产生重要的影响。

在对上海天桥发展的回顾和展望中，我们发现天桥是上海城市建设发展的一个缩影，经历了从建到拆到再建的过程，这正是上海城市功能不断提升、设施不断完善、建设水平不断提高的见证。

天桥，也承载着一代又一代市民的城市记忆，行走在天桥的市民的生活方式和需求逐渐改变，出行的方式也发生了巨大的变化，城市生活越来越丰富，街道活力在慢慢复苏，天桥上的故事和风景也愈来愈精彩。未来天桥将成为城市多元功能融合的桥梁，街区活力链接的平台，立体步行网络的组成部分，城市公共视觉中的一道风景线，成为一个立体、开放、友好、艺术的魅力公共空间。

Note from the Curator

Municipal engineering construction is an important guarantee of the quality of urban spaces. Small factor as it is, the overpass has important implications for the space environment.

A review and outlook of the development of overpass in Shanghai shows that overpass is an epitome of urban construction and development in Shanghai. Having gone through building, pulling down and rebuilding, the overpass has witnessed the improved function, facilities and construction in Shanghai.

The overpass also carries urban memories of citizens from generation to generation. The way of life and demand of those residents walking on the overpass are changing gradually and their means of transportation also changed greatly. City life becomes richer and richer, street dynamism recovers slowly and stories and sceneries on the overpass are increasingly wonderful. Going forward, the overpass will be a multi-functional bridge in the city, a platform linking dynamisms across communities, a component of three-dimensional walking network and a beauty spot in urban public sights. It will become a charming public space that is dimensional, open, friendly and artistic.

策展人 CURATOR

曹晖 Cao Hui
上海营邑城市规划设计股份有限公司副总经理，总工程师。
Vice-manager and chief engineer of Yingba Urban Planning Design Co. Ltd.

庄宇 Zhuang Yu
博士，教授，博导，同济大学建筑城规学院城市更新与设计学科团队主持人，中国城市规划学会城市设计专业委员会委员，国家一级注册建筑师。
Doctor, Professor, PhD Tutor, Host of Urban Regeneration and Design Team at Building and Urban Planning College of Tongji University. Deputy of Urban Design Committee, China Urban Planning Society, Registered Architect Level I

柳亦春 Liu Yichun
大舍建筑设计事务所创始合伙人，国家一级注册建筑师，同济大学建筑城市规划学院和东南大学建筑学院客座教授。
Co-founder of Dashe Architectural Design Firm, Registered Architect Level I. Guest Professor of Building and Urban Planning College of Tongji University and Construction College of Southeast University.

主办单位 SPONSOR

上海市规划和国土资源管理局
Shanghai Municipal Bureau of Planning and Land Resources
上海市文化广播影视管理局
Shanghai Municipal Administration of Culture, Radio, Film and TV
徐汇区人民政府
People's Government of Xuhui District

承办单位 UNDERTAKER

上海西岸开发（集团）有限公司
Shanghai West Bund Development (Group) Co, Ltd
徐家汇街道
Xujiahui Street Community
上海营邑城市规划设计股份有限公司
Yingba Urban Planning Design Co. Ltd.
上海沃华文化传播有限公司
Shanghai Wohua Cultural Transmission Co. Ltd.
同济大学建筑与城规学院 * 城市更新与设计团队 Urban Regeneration and Design Team at Building and Urban Planning College of Tongji University.
上海大舍建筑设计事务所
Dashe Architectural Design Firm
上海索益公益文化发展中心
Shanghai Suoyi Public Culture Development Center

协办单位 SUPPORTER

上海城市公共空间设计促进中心
Shanghai Design & Promotion Center for Urban Public Space
上海市城建档案馆
Shanghai Urban Construction Archives
徐汇区规土局
Xuhui District Planning and Land Administration Bureau
静安区规土局
Jing'an District Planning and Land Administration Bureau
长宁区规土局
Changning District Planning and Land Administration Bureau
闸北区规土局
Zhabei District Planning and Land Administration Bureau
杨浦区建交委
Yangpu District Construction and Transportation Committee
浦东新区建交委
Pudong New Area Construction and Transportation Committee
普陀区规土局
Putuo District Planning and Land Administration Bureau

地点 LOCATION

徐汇区龙腾大道 2555 号 西岸艺术中心
West Bund Art Center, No. 2555, Longteng Avenue, Xuhui District

20世纪80年代徐家汇天桥
Overpass of Xujiahui in 80s'

今天我们开向阳院
静安公共文化艺术实践案例展
Welcome to Our Sunflower Yard
Site Project of Public Art & Culture in Jing'an

经过二十多年的快速发展，静安区在城市建设、经济实力、社会服务体系等方面均取得了较显著的成果。随着城区发展进入成熟期，大规模旧区改造和城市建设也将告一段落。但是，对比世界著名城市的中心区，静安在空间复合度、功能集聚度、服务配套度、交通便捷度等方面还存在一定差距，且还存在着自身发展的诸多困境。一方面，区际间同质化竞争与实现静安差异化发展之间存在一定矛盾；另一方面，大量的存量载体功能有待提升，这其中既包含大量年代久远、服务落后的办公楼宇，也包括闲置工业厂房和利用效率不高的创意园区，更有大面积成片保留和有待置换的历史保护建筑片区，亟需通过对存量资源的二次开发和深度开发，实现城区发展的优化和升级。

城市更新，是针对城市活力衰退的区域进行调适的再活化，从而解决城市问题、促进地区经济环境的可持续发展与社会品质的提升。因此，对于静安区来讲，城市更新是发展的必经之路，是提升城市品质和功能的必然选择，也是集约节约利用土地资源的有效途径。

作为上海城市建设的核心区域，静安区承担着传承城市文脉、更新城市面貌的历史责任。

Having rapidly developed for more than two decades, Jing'an District has made great headways in urban construction, economic strength and social service system. As Shanghai urban development comes to maturity, old area renovation and urban construction on a large scale will come to a conclusion. However, Jing'an still lags behind those central districts of world-famous cities in terms of spatial linkage, function concentration, supporting services and traffic convenience, and also experiences many development difficulties. On the one hand, homogeneous district competition prevents Jing'an from realizing differential growth; on the other hand, functional upgrade is needed for abundant land reserves, including many old office buildings with poor services, obsolete factories and underutilized creative industry centers, and a large area of historical buildings which require protection and renovation.

To promote urban development and upgrade, it is necessary to make secondary, in-depth development of these land reserves.

By moderately revitalizing those declined urban areas, urban renewal helps address urban development issues for a sustainable regional economy and higher social quality. In this light, Jing'an District has to rely on urban renewal to boost local development, improve urban quality and functions, and realize intensive utilization of land resources.

As a key area in Shanghai urban construction, Jing'an District shoulders the historic mission of inheriting local historic culture and updating the look of Shanghai.

左图　利用装置艺术搭建的时空隧道
LEFT　A Time Channel of Installation Art

右图　历史照片中的里弄生活与现今
　　　"空"斯文里
RIGHT　Alley Life in Old Pictures and Today's Empty Siwen Alley

展览主旨

艺术和文化对城市空间的利用激活了城市的创造力，也成为复兴城市文化、重塑城市形象的驱动力，它改变了城市文化的关注和影响轴心。因此，用艺术去"更新"城市，成为关键路径。

展览介绍

公共艺术介入民众生活，是全球公共艺术的整体走势，只有给当地生活带来品质提升的公共艺术才是真正值得倡导的公共艺术。在上海市规划和国土资源管理局的倡导下，以静安区规划和土地管理局为主体，举办了"2015静安城市空间艺术季"系列活动。艺术季不仅面向所谓专业、精英人士，更是渗透到城市日常生活的每个角落，特别强调公众的参与性和实践性。因此，本次静安精心挑选三个独具特色的活动地点：

右页　利用镜面反射形成的装置艺术道
FACING PAGE　Installation Art Road via Specular Reflection

左图	日常生活用品构建的装置艺术
LEFT	Installation using daily objects as materials

"空"斯文里

代表"石库门保护"典型案例的上海最早具有民国建筑风格的联排石库门片区东斯文里。随着静安旧区改造工作基本完成、城市基础设施建设不断推进，对成片里弄住宅进行保护性开发将是"十三五"时期城市更新的重点，这对保留上海城市年轮，传承海派历史文脉，提升现代城市功能，具有十分重要的意义。由此，静安区以里弄空间为载体组织了公共艺术活动——"东斯文里实践案例展"。活动现场，利用里弄空间搭建艺术装置，将传统的里弄文化与当代艺术进行一次有趣的融合，让东斯文里成为一个城市文化和生活不断传承和延续的精神场所；通过艺术家与居民的艺术互动和文化交流，勾起上海人和"新"上海人对里弄文化的回忆和认知，重塑符合时代潮流、与时俱进的里弄文化精神内涵，重新引起社会对里弄文化的关注和探讨。

"乐"里弄音乐

作为"旧区改造"的优秀案例、改造后回迁居民最多的生活社区新福康里。活动以亲民艺术进社区的形式受到热烈欢迎。活动在小区中央绿带内搭建了一个主舞台,舞台周边放置的艺术雕塑及装置与观众席结合,形成生动活泼的观演空间,为居民带来了多元化的音乐文化氛围和视听感受。喜闻乐见的艺术作品与社区居民的零距离接触,提高了艺术参与性,也提升了公共艺术的"公众性"。

"艺"邻里市集

代表"设施改造"中地下泵站结合滨水公共绿地设计最成功的蝴蝶湾公园。这里汇集了多种形式的音乐演出与艺术集市,以"艺术在民间"与"艺术回民间"的形式,让民众在享受艺术的同时,感受到点点滴滴渗入生活的艺术美。这也印证,"静安"已经初步形成了社会主体广泛参与、文化活动丰富多彩、艺术氛围随处可见、广大市民遍地分享的公共文化越来越活跃的整体艺术氛围。

上图　民谣乐队和手鼓艺术家为居民献上一场视听盛宴
ABOVE Performance by folk band and frame drum player

下图　蝴蝶湾公园入口处的艺术集市
BELOW Art Marketplace at the entrance of Butterfly Beach Park

1　手鼓表演艺术家与现场居民的热情互动表演
Passionate interactive Performance between frame drum player and citizens

2　社区居民与装置艺术的欢乐互动
Happy interaction between community residents and installation art

3　居民与人体雕塑合影留念
Residents taking pictures with body sculptures

上图　伴随手鼓表演而翩翩起舞的异域风情舞蹈艺术家

ABOVE Exotic dancing accompanied by frame drum

下图　创意集市让市民欣赏和体验到丰富多彩的创意制作和时尚产品

BELOW Citizens appreciating creative works and fashion products of all kinds at the marketplace

Theme of Case Exhibition

The combination of culture and art with urban space stimulates the creativity of a city, and also drives its cultural revival and image rebuilding. What's more, it changes the focus and influential pivot of urban culture. For this reason, art is no doubt an essential path for urban regeneration.

Exhibition Introduction

The introduction of public art into people's life is a direction where global public art goes. Only public art which improves local living quality are worthy of encouraging. Upon the call of Shanghai Municipal Bureau of Planning and Land Resources, Jing'an District Planning and Land Authority launches a series of activities named "2015 Jing'an Urban Space Art Season". The "Art Season" not only targets at specialists and elites, but aims to penetrate into daily life of Shanghai people, with great importance attached to the participation and practice of the public. Therefore, three unique sites in Jing'an District are chosen:

"Empty" Siwen Alley

East Siwen Alley, an earliest townhouse of the Republican style in Shanghai, is a typical case for "Shikumen Protection". As the old area renovation of Jing'an District comes to an end and the urban infrastructure construction progresses, protective development of massive alley residences will be the focus of urban renewal in the "13th Five-year Plan", which is significant to protect Shanghai's city history and traces, as well as enhancing functions of urban city. Jing'an District Municipality, therefore, held "East Siwen Alley Project Exhibition", a public art event building on alley space. Artistic installations are built with the aid of alley, presenting an interesting combination of the traditional alley culture and modern art, thus making East Siwen Alley a spirit place where urban culture and life are passed on. Through art interaction and culture exchange between artists and residents, the memory and awareness of alley culture are evoked in the mind of local and migrant ones. It shapes the alley culture to fit in the current trend, and attracts social attention and exploration of alley culture.

Music of Alley

The new Fukang Alley, an outstanding case of "old area renovation", welcomes back most residents. The art event receives great popularity. A main stage is set up in the central green belt of the community, and surrounding sculptures and installations blend into the audience seats perfectly, presenting a vivid and vigorous performing and watching space. Diverse music cultures and audiovisual experience is offered to residents through easy access by residents to these popular works of art. This promotes participatory art and makes public art even more public.

Neighborhood Marketplace

Butterfly Beach Park, a successful design combining underground pump station and riverside public green space for "facility transformation", gathers various music performance and art marketplace. By "finding art in the folk" and "bringing art to the folk", the citizens can enjoy the beauty art, and experience art's presence in everyday life. This reflects that Jing'an District has already become a space of increasing public culture participation, featured by extensive participation in social themes, various forms of cultural events, pervasive love for art, and culture sharing by all citizens.

策展人感想

2015上海城市空间艺术季（SUSAS）以打造具有"国际性、公众性、实践性"的城市空间艺术品牌活动为主旨，通过艺术季活动期望达到改善生活空间品质、提升城市魅力的目标。

"公共艺术"体现了公共空间中文化开放、共享、交流的一种精神与价值。公共艺术在城市中的发展，在中国面临着机遇和挑战。"机遇"是城市化发展给公共艺术带来的可能性；"挑战"是如何去选择具有高品质的艺术、艺术品，以适应时代的需要。

在静安，最具代表性的城市空间形态就是里弄。保护和继承里弄文化，是新的历史发展机遇下文化复兴的一种有效途径。而社群公共艺术的公共意义与艺术价值也是值得静安政府认真思考的问题。

艺术如何为静安乃至上海这个城市建构其本土意识和文化认同提供可行的实践路径？我们的"2015静安城市空间艺术季"系列活动，即以此为思考原点，并在此基础上提供了一个有益的实践范本和践行的途径。期望通过公共艺术的探索和创新，为城市空间提供更多表达的可能性。

Note from the Curator

2015 Shanghai Urban Space Art Season (SUSAS) aims to launch International, Public and Practical urban space art events to enhance the living space and charm of Shanghai.

Public Art reflects the spirits and values to open, share and exchange cultures in public spaces. The urban development of public art confronts both opportunities and challenges in China. The opportunities come from the artistic possibilities in urbanization and the challenges lie in how to choose quality art and artworks to fulfill the needs of the times.

Lanes and valleys, the most representative urban spaces in Jing'an, of which cultural protection and inheritance are effective approaches to revive our culture. The public significance and artistic value of community public art are worthy of considerations for Jing'an local authorities.

How does art provide feasible approaches to shape up the local awareness and cultural identity for Jing'an District or Shanghai? This is the point of departure and the basis of 2015 Shanghai Urban Space Art Season to provide a conducive practice model and action channel. We hope to maximize the possible expressing channels for urban spaces by exploring and creating public art.

策展团队 CURATORIAL TEAM
上海罗浮紫艺术品有限公司
Shanghai Purple Roof Artwork Co., Ltd

主办单位 SPONSOR
静安区人民政府
People's Government of Jing'an District

承办单位 UNDERTAKER
静安区规划和土地管理局
Jing'an District Planning and Land Management Bureau

地点 LOCATION
上海市静安区新闸路 568 弄 东斯文里
上海市静安区新闸路 888 弄 新福康里
上海市静安区康定东路 28 号西侧 蝴蝶湾绿地

East Siwen Lane, Alley 568, Xinzha Road, Jing'an District, Shanghai
Xinfukang Lane, Alley 888, Xinzha Road, Jing'an District, Shanghai
Butterfly Bay Green Space, West to No. 28, East Kangding Road, Jing'an District, Shanghai

与市民热情互动的手鼓表演
Drum performance is passionate interaction with the public

承上启下・见微知著
徐汇风貌区保护更新
2015 实践案例展

Transitional From and Subtle Gesture
Conservation and Regeneration in Xuhui Historic Area 2015 Site Project

上海市衡山路－复兴路历史文化风貌区于上海近代"黄金时期"形成，曾是上海法租界西区，至今保留了大批优秀历史保护建筑和历史街道，上海中心城内 64 条一类风貌保护道路中近半数分布在这一区域。

Hengshan Road-Fuxing Road Historical and Cultural Area dates back to the "Golden Age" in Shanghai's modern history. Originally, it was the west district of Shanghai French Concession, leaving us a collection of significant historic buildings and streets. Among 64 historic and cultural streets in Shanghai midtown, almost half is at here.

2015 上海城市空间艺术季的学术委员会组长郑时龄院士这样评价此案例展：

以"承上启下，见微知著"为主题的"徐汇风貌区保护更新 2015"，由策展人和策展团队精心策划和准备，推出了这个既是展览又超越了纯粹展览的综合活动，内容丰富，形式多样。这是一个接地气的活动，既是一个事件，又是一个围绕该片区风貌保护更新长期实际工作中的一个篇章。这个活动与社区居民的日常生活环境息息相关，将艺术家的作品与城市生活场景融为一体，把展示活动融入城市肌理，让市民参与、认识并感受徐汇的历史风貌、历史建筑、城市空间和人文特色。活动的组织安排也和多个承担风貌区日常管理职责的重要部门紧密结合，体现出徐汇各个相关部门正在通过协同配合提升管理能力的意图。除了国际交流内容，这次活动展示的规划实践案例、艺术作品、社区文化培养和传承，以及陈丹燕和尔冬强两位著名文化人士主持的活动，都强烈体现出来自一个具有悠久文化传统的城市区域，并服务于这一社区的特征。

展览主旨

此案例展包含如下四个展览："徐汇风貌道路保护规划与实践 2007–2015"；"魅力衡复·梧桐艺术季——城市更新经典案例展"；"魅力衡复·梧桐艺术季——画家笔下的风貌区"；"世界小学'老洋房探踪'特色拓展课程展览：传承城市文化——我们在武康路上"。通过组织相关讨论和公众活动，引发社会各界对风貌区保护更新的广泛关注，对徐汇风貌区更新发展起到"承上启下"的作用。同时，本次文化活动将体现国际化特点，与国际机构合作，以"见微知著"的艺术理念，将艺术家的作品与城市生活场景融为一体，把展示活动嵌入城市肌理，邀请市民以"慢行慢品"的方式，了解和感受衡复风貌区建筑、城市空间和人文特色。

展览介绍

徐汇区风貌道路保护规划与实践 2007–2015

《徐汇区风貌道路保护规划与实践 2007–2015》展示 2007 到 2015 年的 8 年间，徐汇区在风貌道路保护规划和整治实施方面所做的一系列探索性工作。这些工作在保护规划编制方法、各个部门协同实现精细化管理、发挥规划对近期整治举措的引导和指导作用等方面取得了实效。展览包括三方面主要内容：

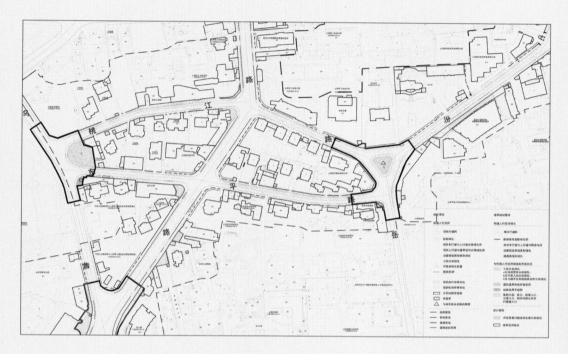

上图　徐汇区风貌保护道路规划 宝庆路—衡山路分册街道平面设计导则

ABOVE "Baoqing Road - Hengshan Road" Historic protection planning of Xuhui District Street graphic design guidelines

1. 武康路——保护规划和整治实施试点 2007–2009

武康路位于上海衡山路—复兴路历史文化风貌区西端，是一类风貌保护道路。该路最初辟筑于 19 世纪末，是近代上海法租界西区内历史最久远的城市道路之一。

自 2007 年初至 2009 年底，作为上海市风貌保护道路保护规划编制的试点，也作为徐汇区政府"迎世博三年行动计划"的重要组成部分，武康路沿线进行了修详层面的保护规划并依据保护规划进行了一系列保护整治工作。由徐汇区历史建筑和历史文化风貌区保护委员会办公室牵头，由总规划师、专家组、规划管理及所有实施部门形成的联席会议模式对确保整治实施与规划合理衔接，确保整治工作合理进展发挥了重要作用。整治中对外露的各类线路产生的不利影响及沿线弄堂内的破败情况等涉及基础设施和居民生活环境方面问题的整治力度很大，"形象"方面的整治主要集中在补种行道树、优化弄堂口部空间和历史建筑所在地块的围墙等小举措，但通过组织不同建筑师参与和设计控制等方式确保多样性和精细化。整治工程避免了焕然一新，但整体上有显著品质提升，同时受居民欢迎。

2. 徐汇风貌道路保护规划 2011–2013

2011 年至 2013 年，基于武康路试点经验，徐汇区对其上海市衡山

147

安福路 南侧

淮海中路（宝庆路—陕西南路） 北侧

长乐路 南侧

新乐路 南侧

襄阳北路

新乐路 北侧

建国西路

高安路 西侧

华亭路

延庆路 北侧

延庆路 南侧

路—复兴路历史文化风貌区内所有道路进行保护规划，在徐汇区规土局组织与协调下，采取总规划师制度，通过与徐汇区房管、历史保护、市容、街道、绿化和市政等相关部门密切配合的工作机制，通过包含规划编制、实施试点和研究提升三方面的一系列探索性工作，将已有的建立在控制性详细规划层面上的保护规划进一步深化为可供政府各相关部门作为日常管理依据的管理指南，使风貌区的保护与管理走向精细化。

规划范围总面积 4.4 平方公里，共有 42 条城市道路（含四界道路），总长度 39.3 公里，其中包含上海市一类风貌保护道路 31 条，几乎占上海全部 64 条一类风貌保护道路总量的一半，是上海中心城内风貌保护道路最密集的区域。为了最大限度实现对风貌区保护规划的深化和细化，规划工作涵盖了该范围内的所有街道。

3. 从规划到实践——徐汇风貌整治行动计划研究 2014–2015

徐汇风貌道路保护规划得到多方面好评，但如何确保规划真正发挥作用，真正切中风貌区目前面临的若干难题，从规划走向能够落地的行动计划，并逐步启动必要的整治提升项目，是目前徐汇区规土、街道、房管、徐房集团等相关部门正在合力开展的实际工作。概括而言，正在筹备和准备实施的行动计划和项目包含近中远期的考虑，包含历史建筑被不合理使用而产生破坏、民生、社区建设、合理引进市场力量、公共参与等多个重要方面，未来三年综合行动的目标是合理启动徐汇风貌区保护更新的进程，提升风貌区综合能级和品质，并探索以新思路和新模式开展工作。

魅力衡复·梧桐艺术季——城市更新经典案例展

城市更新就是生活在城市中的人，对于自己所居住的建筑物、周围的环境或出行、购物、娱乐及其他生活活动有各种不同的期望；对于自己所居住的房屋的修理改造，对于街道、公园、绿地等环境的改善有要求，以形成舒适的生活环境和美丽的市容。上海衡复投资发展有限公司始终秉承"尊重历史、传承文化"的理念，致力于让历史街区重现活力，探索历史建筑的保护利用，发掘历史建筑的文化底蕴。衡复公司修缮改造过的项目有湖南路 285 号（原同盟会李秋君女士上海灾童教养所原址）、岳阳路 190 号（原民国霖生医院旧址）、武康路 40 弄 4 号（原颜福庆旧居）及成片历史街区建业里、衡山坊等。此次展览将向市民展示部分改造项目的成果，在复原建筑本身面貌的同时，让这些历史的记忆永续流传，让历史建筑成为展示海派文化的窗口和载体。

魅力衡复·梧桐艺术季——画家笔下的风貌区

衡山路–复兴路历史风貌保护区是上海市中心城区最大的历史文化风貌保护区，汇聚了两千多幢老房子，是上海最具特色的建筑群落，展现

上图 魅力衡复·梧桐艺术季——画家笔下的风貌区活动现场

ABOVE Charming Hengfu · Chinese Parasol Art Season - the Historic Area under the Pens of Artists

了各国的建筑精华，显现出不同时期的艺术风格，以杰出的构思和精彩的工艺，成为上海富含历史文化财富的魁宝。上海衡复投资发展有限公司始终秉承"尊重历史、传承文化"的理念，深入挖掘历史建筑的人文价值，曾先后参与过《历史建筑水彩画》（上海书店出版社，2010年）、《梧桐树后的老房子》（上海画报出版社，2001年）等画册的编撰工作。此次举办的"画家笔下的风貌区"活动将有一批上海知名艺术家，用钢笔、马克笔、油画棒勾勒出衡复历史风貌区内的历史建筑、特色街区和人文景观。通过绘画活动的展开，全方位、多角度地展示风貌区的建筑特色和人文艺术，呈现一种别具风格的视觉体验。

世界小学"老洋房探踪"特色拓展课程展览，传承城市文化
——我们在武康路上

创立于武康路，至今仍位于武康路的世界小学立足于武康路周边历史街区和历史建筑的人事物情，开发了具有文化和素质教育特色的"老洋房探踪"拓展课程。从城市文化和历史街区保护的角度来看，这个课程的可贵之处还在于——这是来自社区的智慧，显示了徐汇历史文化风貌区内真实的文化底蕴。这个展览呈现世界小学师生自2009年创立该课程以来开展相关活动的一系列"花絮"和思考，表达了传承城市文化的育人理念。展览期间还将举办交流研讨会，讨论基于社区的城市文化传承问题。

上图　世界小学"老洋房探踪"特色拓展课程展览
ABOVE Exploring the Historic Foreign-style Houses, a Special Extension Curriculum of World Primary School

下图　尔冬强谈 SHANGHAI ART DECO
BELOW Deke Erh's Talk of SHANGHAI ART DECO

| 左图 | Art Deco 建筑 |
| LEFT | Art Deco styles |

沙逊大厦，建于 1929 年；(中) 中国银行大厦，建于 1937 年；(右) 国际饭店，建于 1934 年

活动

尔冬强谈 SHANGHAI ART DECO

走入尔冬强的视觉文献，追溯 Shanghai Art Deco 的起源、变迁和再生，解读曾经孕育其萌芽的城市背景以及由它引领出的时代风尚。尔冬强的视觉文献绝非简单的城市影像资料，更是城市文明意识的载体和历史风貌保护的推手。尔冬强对世界范围内 Art Deco 风格的收集和整理更是为理解和传承 Shanghai Art Deco 提供了丰富的参考文献。

陈丹燕讲谈社区记忆是城市更新的生命力所在：
以武康大楼居民口述史为例

什么是城市更新必要的准备？空间整理、规划和发展以什么为标准？社区在城市更新中是怎样的生命体，是被改造和打扮成与现代城市想象相符合的地域，还是寻找和展示自己个体历史与地域文化内涵的社区再造过程？陈丹燕以正在缓慢摸索与推进的湖南街道社区居民口述史为例子，讨论在上海城市更新的规划中，一个成熟的社区能提供的帮助，以及一个文化悠久的社区的期待。

建筑师眼中的徐汇风貌区——风貌整治和老建筑改造设计交流研讨会

随着上海城市发展转型，近年来，越来越多境内外建筑师参与到上海中心城内历史文化风貌区保护整治和老建筑改造工作中。上海的经济社会发展阶段与国际发达城市仍有一定差距，这方面工作目前仍处于起步摸索阶段，在理念、实际情况和操作办法等方面仍有很大改进空间。

左图　建筑师眼中的徐汇风貌区
LEFT　Xuhui Historic Area in the Eyes of Architects

同时，大量风貌区内的整治工作是由政府管理部门牵头，今后很长时间这方面工作仍将重点关注公共利益和民生问题，因此，要求设计环节与管理、社区居民参与和实施环节等方面的紧密结合，这与通常的设计项目在程序上也存在很大差别。目前，国内这方面的可参考案例很少，上海必须探索适合目前实际情况，并切实有效地配合风貌保护整治和老建筑改造设计的操作办法。

在过去的10余年中，徐汇风貌区开展了一系列保护整治探索性工作，很多管理部门、建筑设计单位、实施单位以及社区居民都从不同角度参与了这些工作，也参与了规划和设计方案的讨论，形成了基于徐汇实践的一些经验。参加本次交流会的建筑师都参与过徐汇风貌区保护更新设计的实际工作，各有不同的经验、教训和从各自角度的积极建议。研讨会分享各位建筑师的经验，并与相关管理和实施部门人员形成对话，就今后如何改进设计和实施模式等问题进行交流。

基于社区的城市文化传承——世界小学《老洋房探踪》特色拓展课程交流研讨会

在徐汇风貌区保护更新日益受到社会各界关注的今天，世界小学开展的《老洋房探踪》拓展教育探索与学校真实的历史紧密关联，与所在社区的真实场景紧密联系，是来自社区的、不可多得的一个特殊案例。从教育的视角看，世界小学在老洋房探踪活动中采用的美术、音乐和话剧表演等形式的影响将是深远的，在小学时埋下的种子一定会在学生今后的人生历程中绽放出美丽的花朵。从城市管理角度看，当前实际工作中经常面临对城市保护认同不足，缺乏公共参与意识等问题，世界小学师生的探索，是一种自发的城市文化建设举措，有力地显示了上海这座城市的文化底蕴，是具有徐汇风貌区特色的社区参与，为今后徐汇风貌区保护更新中如何激发社区公共参与，如何建设社区提供了启示。

右图　武康路及周边地区，1947年
RIGHT　Wukang road and the surrounding area, 1947

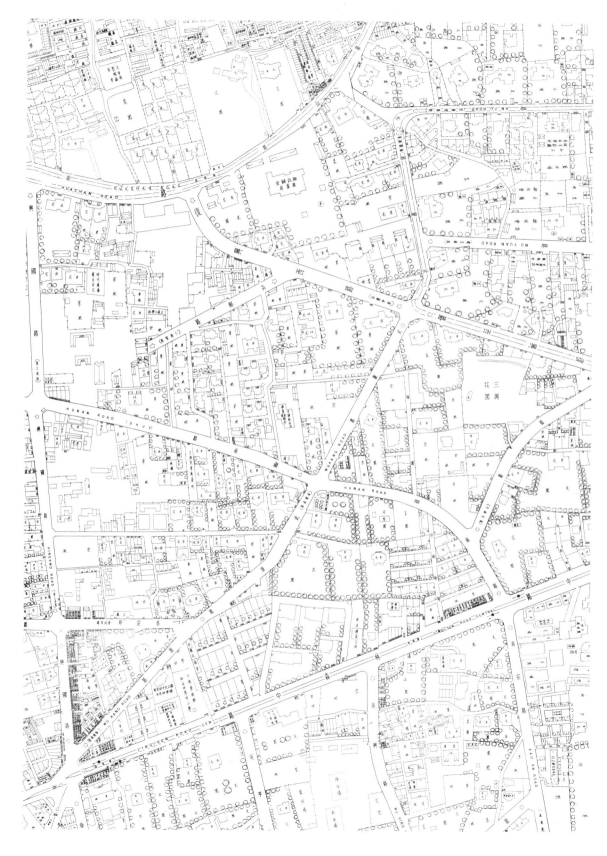

交流研讨会将汇集来自世界小学的师生代表、社区建设和文化相关管理部门、规划设计专业人员，基于世界小学这一特色案例，共同讨论基于社区的城市文化传承问题。

下图　镜圈
BELOW　Mirrored Circles

国际文化交流

见微知著——加拿大公共艺术在徐汇

《见微知著——加拿大公共艺术在徐汇》从武康路出发，结合艺术作品展示和公共文化项目中的对话交流与人文梳理，将外来与本地的文化视角交错穿插，为艺术如何介入生活提出种种新的可能性。这一系列活动的具体内容包括：

艺术作品：《镜圈》、《间联》、《皮橡筋》、《凹园》
1. 作品《镜圈》　艺术家：阿德里安·布莱克威尔
两张镜面不锈钢板上切出的六个同心圆，制成公共街具：座椅、圆桌、壁架和镜子，分散在兴国绿地的不同位置。它们的反射面将周围环境中的不同元素拼接成出其不意的并置画面，同时它们的形态也为不同大小的人群排列了各种就坐和互动的组合方式。

2. 作品二：《间联》　艺术家：内斯特·克鲁格
《间联》取自一种常见的六根型三轴组合积木，历史出处，说法不一，其中一种就是鲁班锁。积木的组合原理是六根条棍插在一起组成一个立体十字结构，整体结构的中间是实心的。六根积木组件周身的图案展开铺平，连接起来也是一幅拼图，一幅武康路及周边相连街道的地形图。

3. 作品三：《皮相筋》　艺术家：柯乔
艺术家以司空见惯的橡皮筋为创作原点，貌不惊人的日常生活细节经过扫描、放大、铸型、上色等一系列处理后，变身为形态生动的现代抽象雕塑。作品提醒我们对于日常的关注与发现，更新未必需要无中生有，可能性可以就在身边。

4. 作品四：《凹园》　艺术家：刘任钧
凹园是一组几何形状的的下沉式流动花坛，周身的印花取历史保护建筑内的瓷砖图案以及反渗出的木板纹样，花坛内会种植上海本地草本植物。凹园被打散，分别置放在两家坐落在与武康路相接道路上的艺术空间内：东画廊（复兴东路），Big Space（兴国路）。

公共文化项目：徐汇风貌区"自行游"之"消失的街坊"和"文人笔下的空间"
展览期间定期通过微信平台推送可供下载的视频和音频导览内容，鼓励游览者自行安排组织行走路线，了解和感受徐汇风貌区人文与社会

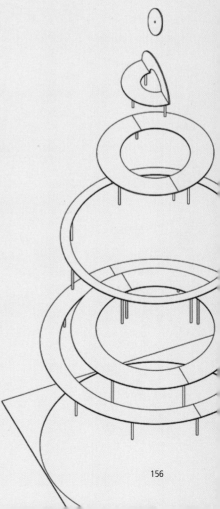

空间的形成与变迁。通过对散落在民间的人文空间史的收集和整理，配以曾在文人笔下出现过的对风貌区空间描述的文字，让游览者浸入式地体验徐汇风貌区老房子背后的日常生活，从而了解风貌区的更新究竟新在何处。

消失的街坊，是一组微访谈影像记录系列。受访对象为武康路及周边地区曾经和现有的住户，分别介绍各自记忆中该地区的生活空间（建筑、社区、生活方式等），重访原址。通过受访者的描述与采访发生时该地点现状的对比折射出武康路及周边地区在城市更新过程中生活空间的变迁。文人笔下的空间，则梳理中外作者对徐汇风貌区空间描述的文字，精选一组制作成语音版，游客边走边听，通过对实际空间多元化的体验，感受文人笔下的徐汇风貌区。

中加对话之一：从艺术与公共空间谈起
现场邀请了加拿大多伦多大学美术馆馆长芭芭拉·菲舍尔女士，上海独立策展人、乌拉尔当代艺术工业双年展策展人比利安娜·思瑞克女士，多伦多艺术家、加拿大圭尔夫大学艺术音乐学院教授柯乔先生，上海艺术家、收藏家、策展人施勇先生，由策展人吴彦女士主持，以一对一访问的形式，配对中加艺术工作者，探讨城市更新进程中的艺术创作以及艺术与公共空间的互动关系，试图通过加拿大视角来重新认识和解读上海的城市现象和艺术实践。

中加对话之二：风貌保护与更新改造
邀请中加建筑与规划领域的学者、建筑师、设计师和规划师从发生在加拿大和徐汇风貌区的实际案例出发，讨论风貌保护的方法手段，实施成效，以及下一步城市更新中风貌保护的未来。

意大利主题文化展 —— 共享的美好食光
"民以食为天"是中国的一句古话。事实上，饮食在世界各地普遍都得到人们的高度重视。在欧洲的修道院文化中，餐室总是一幢建筑中最受欢迎、最能体现审美艺术的场所。米兰的圣玛利亚修道院就是一个绝佳的例子，它因餐室而举世闻名。餐室本身的设计极具美感，达芬奇的《最后的晚餐》和乔瓦尼·多纳多的《十字架》也作为壁画收藏于此。

近年来，饮食空间的设计重新引起了设计师们的重视，他们再次聚焦餐室的审美、环境和功能等内容。借助 2015 年米兰世博会的主题 "Feeding the Planet, Energy for Life"，米兰理工大学与同济大学共同举办了这次展会，探讨餐饮空间的设计问题，同时助力推广世博会的理念，传播相关领域的前卫设计。此次展览以"共享的美好食光"为题，介绍了餐饮空间的演化历史，介绍优秀的餐饮空间设计作品，呼吁为人们创造更高品质的饮食环境。

Said academician Zheng Shiling, academic committee leader of Shanghai Urban Space Art Season 2015 for the case exhibition:

With the theme of connecting link and subtle gesture, the 2015 Xuhui Historic Sights Area Protection and Update Exhibition is elaborately planned and prepared by the planners and their team. This comprehensive event is more than an exhibition featuring exuberant contents and diverse formats. It is an event showing care for the public, as well as a chapter aligning with the long-term practice in the saturation update of the historic sights area. This event is closely linked with the daily life environment of the community residents, and combines the artists' creations and the urban living settings, so that the exhibition is deep inside the city, allowing the citizens to access, recognize and perceive the historic sights, historic buildings, urban space and cultural characteristics of Xuhui. The organization of the event is also based on the cooperation with several major departments responsible for the daily management of the historic sights area, which reflects the Xuhui authorities' intent to enhance the management capability by cross-functional collaboration. Besides the international exchange, the displayed planning practice cases, artistic creations, the community culture foster and heritage, and the activity chaired by the two famous cultural celebrities Chen Danyan and ErDongqiang, all these are conveying a clear message: it originates from an urban area of long-standing cultural heritage, and serves the community.

Theme of Case Exhibition

This exhibition seeks to trigger a wide focus from public through discussions and activities, making it a turning point for updates and developments of Xuhui Historic Area. Cooperated with international organizations, this cultural activity presents the art concept of "Recognizing the Whole Through a Part" and integrates the artists' work/display activities and urban scenes of life into a whole. Here, citizens are invited to taste and experience the architectures, urban space and cultural characteristics of Hengshan Road-Fuxing Road Historical and Cultural Area in a slow way.

Exhibition introduction

Planning and Practice of Xuhui Historic Road Protection 2007–2015

"Planning and Practice of Xuhui Historic Road Protection 2007–2015" presents the exploratory endeavors made by Xuhui District in planning and managing historic road protection from 2007 to 2015. During these 8 years, much fruits have been born in preparing protection plan, in delicacy management through cross-department collaboration, and in guiding recent managements through planning. This exhibition comprises the three aspects below:

1. Wukang Road — protection planning and rectification implementation pilot project from 2007 to 2009
Wukang Road is a class I historic sights protection road located at the west part of the Hengshan Road — Fuxing Road historic and cultural area. Established in late 19th century, it is one of the oldest urban roads in the west area of the French concession of modern Shanghai.

From early 2007 to the end of 2009, as a pilot project of Shanghai historic sights protection road protection planning, and an integral part of the three-year action plan before the Expo by the

	1			
	2		4	
	3			

1&2 艺术家阿德里安·布莱克威尔作品《镜圈》
"Mirrored Circles" by Adrian Blackwell

3 艺术家刘任钧作品《凹园》
"Recessed Gardens" by Liu Renyun

4 艺术家作品内斯特·克鲁格作品《间联》
"An Interval Connection" by Nestor Kruger

Xuhui district government, the protection plan was given along Wukang Road. Much protection and rectification is carried out as per the protection plan. Led by Xuhui historic building and historic cultural area protection committee office, the joint council approach composed of the chief planner, the expert team, Planning Management and all the execution departments, plays a key role in ensuring the reasonable linkage between rectification implementation and planning, as well as the acceptable progress of the rectification. The project exerts great efforts on the rectification of infrastructure and the citizens' living environment issues, such as the naked lines and cables causing negative impacts and the ruined lanes along the road. The image improvement work mainly includes planting trees, optimized entrance space of lanes, and the enclosure of the plot where historic buildings are located among other small measures. By organizing different architects to participate in and design control, however, diversity and refinement are ensured. The rectification project is not for a brand new look, but for a significant overall quality enhancement, which is popular with the residents.

2. Xuhui historic sights road protection planning for 2011 to 2013
From 2011 to 2013, on the basis of the experience of Wukang Road pilot project, Xuhui district made protection planning for all the roads within the Shanghai Hengshan Road — Fuxing Road historic cultural area. Under the organization and coordination by Xuhui district land planning bureau, and adopting the chief planner system, as well as the working mechanism of close collaboration with Xuhui district housing administration, historic culture protection, urban management, streets, gardening, and municipal administration authorities; through three aspects of many exploring efforts covering planning development, pilot implementation and study for improvement, the existing protection planning based on the detailed control planning is further optimized as a guidance for all the government authorities in their daily management practice, leading to more refined protection and management of the historic sights area.

The planning scope covers a total area of 4.4 square kilometers, with total 42 urban roads (including the four borders) of a total length of 39.3 kilometers; of which, 31 are class I historic sights protection roads in Shanghai, which accounts for nearly a half of the total 64 class I historic sights protection roads in Shanghai, hence it is the area of the most historic sights protection roads within Shanghai central urban area. In order to realize the in-depth and refinement of the protection planning for the historic sights area to the most extent, the planning covers all the streets within the scope.

3. From planning to practice — Xuhui historic sights area rectification action plan study from 2014 to 2015
The protection planning for Xuhui historic sights roads protection is positively recognized; however, it is the current practical work jointly worked by Xuhui land planning, streets, housing administration and Xufang Group to ensure the real effect of the planning, eliminating the several challenges of the historic sights area, transforming the planning into feasible action plans, and gradually introducing the necessary rectification enhancement project. To sum up, the action plans and projects ready for implementation include short-, mid- and long-term considerations, covering the damage to historic buildings due to improper use, people's livelihood, the community construction, appropriate introduce of market force and public engagement among other key aspects. The overall target for the upcoming three-year comprehensive actions is to reasonably launch the process of the protection update of Xuhui historic sights area, improve the overall profile and quality of the historic sights area, and explore new thinking and modes of working.

Charming Hengfu · Chinese Parasol Art Season - Exhibition of Classic City Renewal Cases

Urban regeneration reflects people's divergent expectations for the city where they live - residential buildings, surrounding environment, travel, shopping and all other living activities, and mirrors people's desires for a comfortable environment and beautiful city appearance - by altering their living houses and improving streets, parks and green lands. Shanghai Hengfu Investment & Development Co., Ltd. ("Hengfu") has always adhered to the corporate idea of "respecting history and inheriting culture" and devoted to re-invigorating historic blocks, and exploring cultures, protection and utilization of historic buildings. By far, Hengfu has renovated and renewed many projects, like former Shanghai Home for Child Refugees built by Ms. Li Qiujun in former Animal Defense League (No.285, Hunan Road), former Linsheng Hospital (No.190, Yueyang Road), Former Residence of Yan Fuqing (No.4, Lane 40, Wukang Road) and historic blocks (e.g., Jianyeli and Hengshanfang). This exhibition will demonstrate to the public part of the renovation achievements that restore the original appearances of historic buildings. It also serves the purpose of continuing these historic memories and making the historic buildings as a window and carrier of Shanghai Culture.

上图 意大利主题文化展——共享美好食光
ABOVE Italian Thematic Cultural Exhibition — Shared Happy Time

Charming Hengfu · Chinese Parasol Art Season - the Historic Area under the Pens of Artists

As the largest historic and cultural protection area in central urban area of Shanghai, Hengshan Road-Fuxing Road Historic and Cultural Area is the most characteristic building group in Shanghai converging over two thousands of old houses in all building and art styles spanning different countries and ages. The splendid ideas and meticulous crafts make it a treasure of Shanghai history and culture. Adhering to ideas of "respecting history and inheriting culture", Hengfu has long been exploring culture values behind the historic buildings and engaged in compiling albums of painting like Aquarelle of Historic Buildings (Shanghai Bookstore Publishing House, 2010), The Old Houses behind Phoenix Trees (Shanghai pictorial press, 2001). In "Historic Area under the Pens of Artists", visitors will appreciate the historic buildings, characteristic blocks and cultural landscapes in the historic areas renovated by Hengfu, under the pens, mark pens and oil pastels of famous artists. It will offer a distinctive visual experience for building features and cultural arts of the historic areas in multiple perspectives.

Exploring the Historic Foreign-style Houses, a Special Extension Curriculum of World Primary School - We are in Wukang Road

Seated in Wukang Road since its establishment, World Primary School develops a special extension curriculum named "Exploring the Historic Foreign-style Houses" featured with culture and quality education characteristics based on the history and culture in nearby historic blocks and buildings. This curriculum is valuable for protection of urban cultures and historic blocks in that it is a collection of community wisdoms and reflects real cultural heritages in Xuhui Historic Area. This exhibition showcases a range of "sidelights" and thoughts from students and teachers in performing related activities since setting up of this curriculum in 2009, delivering education concepts of maintaining cultural ties. During exhibition, seminars will be held to discuss community-based urban cultural heritage.

Events

Deke Erh's Talk of SHANGHAI ART DECO

Through Deke Erh's visual literatures presented in this activity, participants will trace the origin, change and regeneration of Shanghai Art Deco, and interpret the urban backgrounds behind and the custom of the times it leads. Those visual literatures are not just simple urban image data, but a carrier of urban civilization consciousness and a drive of historic style protection. Deke Erh's collection and collation of Art Deco styles across the world provide an abundant of references for understanding and inheriting Shanghai Art Deco.

Chen Danyan Talks about Community Memory - Vitality of Urban Regeneration: from oral history of Wukang Building residents

What are necessary preparations for urban regeneration? What are the standards of spatial layout, planning and development? What's the role of community in urban regeneration? Will communities be renovated and shaped into areas adapting to modern city imaginary? Or, will urban regeneration be a communication reconstruction process to seek for and display individual history and regional cultural connotation? Taking the oral history from Hunan Street Community residents as example, which is under slow exploration and progress, Chen Danyan will discuss what a mature community can help in planning Shanghai urban regeneration and what are the expectations from such a community with deep-rooted culture.

Xuhui Historic Area in the Eyes of Architects - Seminar on Style Renovation and Old Building Reconstruction

With urban development transformation of Shanghai, in recent years, more and more domestic and foreign architects have participated in the style renovation and old building reconstruction of the historic and cultural areas in central city of Shanghai. Given the gap between Shanghai and international developed cities in social and economic development, the style renovation and old building reconstruction in Shanghai, still in its infancy, needs much exploration and improvement in ideas, actual conditions and implementation methods. In addition, the renovation of many historic areas is led by government authorities and its key focus will be public interest and people's livelihood for a long time to come. Therefore, different from common design project, close link is to be built between design and administration and community residents are encouraged to participate in project implementation. By far, there are few reference cases in China and Shanghai has to explore feasible and effective implementation methods suitable for current conditions in coordinating with style protection/renovation and old building reconstruction design.

Over the past 10 years, explorations have been made by Xuhui Historic Area about renovation and protection, and plans and design schemes have been discussed, with a wide participation by administration authorities, building and design institutions, construction units and community residents from different aspects to form some experience based on actual conditions of Xuhui Historic Area. All architects present in this Seminar have been involved in protection renewal design of Xuhui Historic Area, and are expected to propose active suggestions from different aspects based on their experience and lessons. In this Seminar, architects will share their experience and dialogue with related administration and implementation authorities to communicate about how to improve designs and implementation modes.

The Community-based Inheritance of Urban Culture - Exchange Seminar of Exploring the Historic Foreign-style Houses - Special Extension Curriculum of World Primary School

Today, among growing focus on protection and renewal of Xuhui Historic Area, the extension curriculum of Exploring the Historic Foreign-style Houses is a rare community-based case that explores extension education and combines school history with real community scenes. The various class forms (e.g., painting, music and opera performance) have long-lasting and deep education significance and what students' learn in this curriculum will exert a tremendous influence on their future. Given a dearth of senses of identity and participation among public in urban protection, the exploration from teachers and students in World Primary School is a spontaneous measure for urban culture construction and represents community participation of Xuhui Historic Area's characteristics, representing the deep culture of Shanghai. It provides important insights about how to aspire community in future protection and renewal of Xuhui Historic Area and public participation and how to construct communities. In this Seminar, teacher and student representatives from World Primary School, community construction and culture authorities, professional planners and designers will gather together to discuss the community-based inheritance of urban culture based on this special case.

International Cultural Exchange Activities

1. Recognizing the Whole Through a Part - Canadian Public Art in Xuhui

With Wukang Road as starting point, this exhibition will explore new possibilities to integrate arts to lives by fusing foreign and domestic cultures with a combination of dialogue communication and cultural sorting in artistic works display and public cultural projects. Four experienced Canadian artists — Adrian Blackwell, James Carl, Nestor Kruger and Yam Lau are invited to create a series of visual works based on the social, historical and space characteristics of Wukang Road area, which are scattered along Wukang Road and within the connected blocks as temporary public places. The subtle gesture is a sort of design attitude that values detail, precision and throughout. During the Art Season, the 4 groups of 10 creations are deployed along Wukang Road and at various public space of the surrounding blocks, including Wukang Road flowerbed, Ba Jin's former residence, Xingguo greenbelt and two art space near Wukang Road — Don Gallery & Big Space. The activity comprises:

Artistic works: Mirrored Circles, An Interval Connection, Thing's End, Recessed Gardens.

Title: *Mirrored Circles*　　By: Adrian Blackwell
Six concentric circles cut from two mirror surface stainless steel plates are made into public settings: chairs, round tables, wall-projected shelves and mirrors, which are scattered at different locations of Xingguo greenbelt. Their mirror surfaces integrate the various elements from the surrounding environment into surprising synchronous images, meanwhile their shapes offer varied combination choices for different people to sit and interact.

Title: *An Interval Connection*　　By: Nestor Kruger
An Interval Connection is inspired by a kind of common toy brick with six sticks and three axles. Its historical origination is not clear, but some said is Luban Lock. The six sticks together can build

a three-dimensional structure of cross. The whole structure is stuffed in the core. The patterns on each of the six bricks, when being put together, can also form a picture — the geographical map of Wukang Road and the surrounding blocks.

Title: *Thing's End - China Red No. 1*   By: James Carl

The artist gets the inspiration from the common bungee. After scanning, magnifying, casting mould and coloring, the seeming plain daily life detail turns into the vivid modern abstract sculpture. The creation calls for our attention and discovery to the daily life. The update is not necessarily out of nothing; the possibility may be just here with you.

Title: *Recessed Gardens*   By: Yam Lau

Recessed Gardens are a group of mobile geometric sunken flowerbeds. The patterns on them are sourced from those on the ceramic tiles of the historic protection buildings and the wood board patterns. In the flowerbeds, Shanghai local plants are growing. The flowerbeds are scattered and deployed in two art space respectively on the roads connecting with Wukang Road — Don Gallery on East Fuxing Road & Big Space on Xingguo Road.

Public area project: *DIY Tour* in Xuhui historic sights area — *The Lost Block and The Space in Writers' Eyes*

During the exhibition, vids and audio guidance contents are regularly sent through WeChat for download, visitors are encouraged to arrange and organize their own DIY tour route, to learn and perceive the forming and evolution of the cultural and social space in Xuhui historic sights area. By collecting and arranging the cultural space history from the folk, plus the writers' words describing the historic sights area space, the visitors are allowed to well experience the daily life behind the old houses in Xuhui historic sights area, and learn the update of the historic sights area.

The lost block is a series of micro interview video records. The interviewees are the former and existing residents on Wukang Road and in its surrounding area, who recall the living space of the area (the architecture, community and life style, etc.) and revisit the former address. The interviewees' description, compared with the address' status quo, reflects the evolution of the living space of Wukang Road and its surrounding area in the process of urban update. *The space in writers' eyes* exhibits the words by the domestic and overseas writers describing the space of Xuhui historic sights area. The audio version of selected collections is available for the visitors to listen as they enjoy the tour, thus feel Xuhui historic sights area in writers' eyes by the diversified experience of the real space.

Sino-Canada dialogue I: start from art and public space

Guests including Ms. Barbara Fischer, director of the art gallery of University of Toronto; Ms. BiljanaCiric, Shanghai independent exhibition planner, planner of the 2nd Ural Industrial Biennial of Contemporary Art; Mr. James Carl, Torontoartist, professor of the school of art and music, University of Guelph; and Mr. Shi Yong, Shanghai artist, collector, exhibition planner are invited to the site. Chaired by the exhibition planner Ms. Wu Yan, Chinese and Canadian artists are paired to receive the one-to-one interview discussing the artistic creation in the process of urban update and the interactive relationship between art and public space, and trying to re-understand and interpret the urban phenomenon and art practice in Shanghai from a Canadian perspective.

Sino-Canada dialogue II: historic sights protection and update modification

Chinese and Canadian scholars, architects, designers and planners in the architecture and planning fields are invited to, based on the real cases in Canada and Xuhui historic sights area, discuss the methods and the implementation results of historic sights protection, and the future historic sights protection in the next urban update.

2. Italian Thematic Cultural Exhibition - Shared Happy Time

As an old Chinese saying goes, "Hunger breeds discontent". In fact, food and drink are highly valued by all the people across the world. In European monastery culture, a dining room is always the most welcomed place in one building, which reflects art of aesthetics to the maximum. A good example is Santa Maria delle Grazie in Milan, renowned for its crafted dining room, featuring aesthetic design and precious murals - the masterpieces of The Last Supper by Davinci and Crucifixion by Giovanni Donato.

In recent years, designers have renewed their interests in dining space and attached importance to aesthetic environment and function in dining room. Based on the theme of Milan Expo 2015 "Feeding the Planet, Energy for Life", this exhibition is coordinated by Politecnico di Milano and Tongji University to discuss design of dining space and to spread flash-forward designs in this field while promoting Expo ideas. Themed at "Shared Happy Time", this exhibition introduces evolutionary history of dining space and presents fine arts to encourage people to create high-quality dining space.

策展人感想

沙永杰

回顾案例展"徐汇风貌区保护更新 2015"过去三个月中的 10 项活动内容，深感当初把这个活动当作风貌区保护更新长期实际工作中的一个环节的组织策略是正确的——活动对今后实际工作有实实在在的推进作用。80% 的活动内容和参加人员直接来自于社区，从世界小学小学生的老洋房特色教学内容，到在武康大楼住了半个多世纪老人家参加的口述史项目，充分显示了社区的底蕴，加上多个管理部门联手组织和参与，实现了"来自社区，服务于社区"的初衷。

Note from the Curator

Sha Yongjie

When looking back to the 10 activity contents of the 2015 Xuhui Historic Sights Area Protection and Update Exhibition in the past three months, the strategy of organizing the event as a part of the long-term real practice in the protection and update of the historic sights area, is definitely right— the event has a practical driving effect on the future practice. 80 percent of the activity contents and the participants come directly from the community. From the unique teaching contents of the old foreign style house in the Shijie Primary School to the oral history project held in the home of a senior resident who has been living in Wukang Building for over half a century, demonstrating the cultural heritage of the community; plus the joint organization and engagement by multiple management authorities, which realizes the original intention of from the community and serve the community.

吴彦

对于城市空间的理解不该停留在静态的规划和建筑,动态的人群和生动的日常才是它的关键。艺术季实践案例展的目的是直接进入社区的日常生活,将展示内容从展场静止的再现中抽离出来,重新置入日常的节奏和脉动。此时,艺术与观众之间的关系不再是被动的观赏,而是主动的体验。然而,城市空间里的艺术体验应该试图远离教条式的灌输和强加的互动,由日常活动中不经意间带动的视觉生产和思考同样可以激活闲置的公共空间,为社区带来积极的活力和可能性。

Wu Yan

The understanding of urban space should not be remained in static plans and architectures, and its key lies in dynamic population and vivid daily life. Our vision is to directly get into daily life of communities and draw the display contents from reproduction of static exhibition hall and reintroduce daily rhythm and impulse. In this way, audiences no longer passively appreciate the arts but enjoy an active experience. Therefore, attempts are required to get art experience of urban space away from dogmatic preaching and imposed interaction. Inadvertent visual production and thoughts in daily life also activate idle public space and pour in active vitalities and possibilities to communities.

策展人 CURATOR

沙永杰 Sha Yongjie
同济大学建筑与城市规划学院 教授
同济大学建筑与城市空间研究所 副所长
Professor of College of Architecture and Urban Planning, Tongji University
Deputy Director of Institute of Architecture & Urban Space, Tongji University

吴彦 Wu Yan
多伦多大学建筑景观和设计学院视觉系策展专业硕士,第五届深圳城市建筑双年展加拿大馆的联合策展人
Master of Department of Visual Studies, Faculty of Architecture, Landscape and Design, University of Toronto; Co-curator of Canada Pavilion, 5th Bi-city Biennale of Urbanism/Architecture in Shenzhen.

芭芭拉·菲舍尔 (Barbara Fischer)
——艺术指导
多伦多 Justina M. Barnicke 美术馆和多伦多大学美术馆的执行馆长和主策展人,多伦多大学建筑景观和设计学院视觉系策展专业主任和研究生部的高级讲师
Executive director/chief curator of the Justina M. Barnicke Gallery and the Gallery in University of Toronto; Director of Curation Major and Senior Lecturer of Graduate Faculty, Department of Visual Studies, Faculty of Architecture, Landscape and Design, University of Toronto.

主办单位 SPONSOR

徐汇区人民政府
People's Government of Xuhui District

承办单位 UNDERTAKER

徐汇区规划和土地管理局 | 湖南路街道办事处 | 徐房集团 | 同济大学建筑与城市空间研究所
Xuhui District Planning and Land Administration Bureau | Sub-district Office of Hunan Road | Xufang Group | Institute of Urban Planning & Architecture, Tongji University

协办单位 SUPPORTER

徐汇区住房保障和房屋管理局 | 徐汇区文化局 | 徐汇区旅游局（徐汇老房子艺术中心）| 徐汇区绿化署 | 世界小学 | 加拿大驻沪总领事馆
Xuhui District Housing Administration Bureau | Xuhui Bureau of Culture | Xuhui Bureau of Tourism (Xuhui Old House Art Center) | Xuhui District Afforestation Administration Bureau | World Primary School | Canadian Consulate General in Shanghai

国际合作单位 CO-OPERATION

加拿大多伦多大学美术馆
University of Toronto Art Centre

策划执行团队 PLAN EXECUTION TEAM

徐汇区规划和土地管理局 朱婷 陈扬 刘阳
湖南路街道办事处 李侃 蔡玮
徐房集团 丁曙 朱劲松 魏敏杰
Xuhui District Planning and Land Administration Bureau (Zhu Ting, Chen Yang, Liu Yang);
Sub-district Office of Hunan Road (Li Kan, Cai Wei);
Xufang Group (Ding Shu, Zhu Jingsong, Wei Minjie)

特别致谢 SPECIAL THANKS TO

陈丹燕与湖南街道口述史项目 | 尔冬强
Chen Danyan & "Spoken History of Hunan Street Community" program | Deke Erh

地点 LOCATION

武康路及周边区域
Wukang Road and surroundings

文中图片均由徐汇区规划和土地管理局提供
All images in this case are provided by courtesy of Xuhui District Planning and Land Administration Bureau

艺术家柯桥作品《皮相筋》
Thing's End by James Carl

空间漫步
"社区与技术"生活业态实践案例展

Space Tour
Site Project of "Community & Technology" Lifestyle

华鑫园区原为漕河泾东区电子制造业聚集地，随着产业升级改造，大量工业地块闲置，华鑫股份有限公司通过对工业地块二次开发，以城市更新的形式，打造新型华鑫高端商务园区，成为上海市城市更新的典范企业。

华鑫园区在作为2015上海城市空间艺术季的案例进行展示的同时，也作为其分会场之一，以街区内公共空间的漫步体验活动为主导，制定整个区域内的参观"空间漫步"路线及各项活动。本次活动由上海华鑫股份有限公司承办，由马卫东、刘宇扬、王家浩共同策展。此外，本次活动还与虹梅庭公益服务中心联手定期组织社区及园区公共活动。

Established in 1984, Caohejing Development Zone is one of the first national economic and technological development zones and it has made brilliant achievements in scientific planning, intensive development, high-tech and socioeconomic benefits. Currently, the development zone is actively promoting the creation of ecological industrial park with circular economy and increasing efforts to explore humanities and arts in urban industrial parks.

China Fortune Park was originally an electronics manufacturing hub in the CHJ east zone. With industrial upgrade and transformation, a large amount of industrial plots are left unused. China Fortune Group, through second development of industrial plot, has created a new China Fortune high-end business park by urban regeneration and became an exemplary industry in this field of Shanghai.

China Fortune Park is both a case for exhibition and one of the parallel sessions of Shanghai Urban Space Art Season 2015. The Space Tour route and all the activities are developed for the whole area in the idea of experiencing a space tour in the public spaces of the block. This event is undertaken by Shanghai Huaxin Holdings Co., Ltd. with Ma Weidong, Liu Yuyang and Wang Jiahao as the co-curators. In addition, Hongmeiting Public Service Center has been invited to regularly co-organize public activities for the community and the park.

1 华鑫展示中心 2 华鑫中心 3 华鑫天地 4 华鑫会议中心
a 华鑫中心（中庭）b 各地美食 c 腾讯创业基地 d 芳草地画廊 e 咖啡馆 f 面馆 g 沿街商铺 h 街角公园
i INESA 产品体验中心 j 平台景观 1 k 水岸景观 l 咖啡点 m 平台景观 2 P 停车场
1 Huaxin exhibition center 2 Huaxin center 3 Huaxin world 4 Huaxin convention center
a Huaxin center (middle courtyard b various cuisines c Tencent start-up base d Parkview Green art gallery
e cafe f pasta g stores along the street h street corner park
i INESA product experience center j sightseeing platform 1 k water shore landscape l cafe m sightseeing platform2
P parking lot

上图　漫游地图
ABOVE Tour map

展览主旨

本次活动主要通过"空间漫步"的现场体验，重新认识新型社区的成长与发展过程，关注社区中的人，体验在街区中的搭建，重新认识新型技术在社区营造中的作用，并激发出"新艺术"形态。

同时，展开社区内的田野工作，挖掘面向新型社区组织的事例重新组织叙事，并配合社区近年发展与建造的成果展，强化新型社区的整体氛围。另外，确定参观者在整个地区"空间漫步"的路径，让参观者对街区进行考察，邀请建筑师和设计师进行实地导览，并配合街区完善公共空间的导示系统建设。另外还邀请建筑师与艺术家的组合团队，对空间漫步路径中的重点场所与空间，进行定制式的创作，形成场景化的艺术与演剧作品。这一空间作品将作为未来社区的公共演剧中心的舞台，定期组织社区的公共活动,以期能够在本案例展所在地形成一个长效的青年文化推广机制。

展览介绍

在从工业主导的经济模式向服务主导的经济模式转变的背景下,华鑫股份经过五年的时间,在漕河泾片区进行了一系列城市更新的实践,取得了丰硕的成果。本次展览作为上海城市空间艺术季华鑫园区"社区与技术"生活业态实践艺术展中三个展览——城市再生实践展、智慧城市展及4号地块文献展,展示了华鑫股份在城市更新实践中诸多的尝试和理念。它既是华鑫园区"社区与技术"生活业态实践艺术展的开幕展,也揭开了华鑫股份在诸多城市更新实践中的序章。此外,两个艺术装置展——"华鑫园区装置艺术——韧山水"和"华鑫天地6+1装置艺术展"也通过艺术介入空间的方式来激活空间。

华鑫城市再生实践展

展览由漕河泾片区总体规划展区、四个独立项目展厅以及漕东片区(四号地块)规划设计国际竞赛方案展区三个部分组成。

四个独立的展厅,通过详尽的图文资料和实体模型,分别向公众展示了来自长谷川逸子建筑计画工房、山水秀建筑事务所、大舍建筑设计事务所以及法国雅克·费尔叶建筑事务所的项目。同时也展示了来自SANAA、David Chipperfield、GMP三家世界顶尖设计事务所的竞赛方案。

智慧城市展

展览由欧洲著名的智慧城市专家Raoul Bunschoten教授与其带领的柏林工大可持续城市研究室策划,共分为两个展厅:展厅1将以"智慧城市六讲"为主要呈现方式;展厅2将以德国柏林智慧城市为案例,畅想其在"智慧城市"发展下的可能性。

Raoul Bunschoten教授是英国CHORA创始人,国际著名智慧城市专家,德国建设部聘请的城市可持续性发展及能源效率顾问,以及中国国务院发改委外籍专家。Raoul Bunschoten教授和他的团队曾经为英国伦敦提供泰晤士河东"智慧城市"方案,获得柏林滕伯尔霍夫机场城市改造竞赛首奖,也曾受邀参加深圳双年展,为厦门规划局在2011年成都双年展上提供智慧城市装置模型。

4号地块文献展

位于上海市市中心西南部漕河泾开发区东区的"4号地块",在国家"十二五"产业发展规划的指引下,正通过自身产业发展转型升级,着力

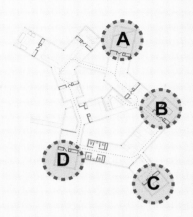

A—— 长谷川逸子建筑计画工房展区
　　 Itsuko Hasegawa Atelier exhibition area
B—— 法国雅克·费尔叶建筑事务所展区
　　 Jacques Ferrier Architectures exhibition area
C—— 大舍建筑设计事务所展区
　　 Atelier Deshaus exhibition area
D—— 山水秀建筑事务所展区
　　 Scenic Architecture Office exhibition area

上图　华鑫展示中心二层平面布局
ABOVE　Plan of the 2nd floor, Huaxin exhibition center

上图　法国雅克·费尔叶建筑事务所展区 / 项目——华鑫天地·漕河泾
ABOVE Jacques Ferrier Architectures exhibition area/project - Huaxin World, Caohejing

下图　长谷川逸子建筑计画工房展区 / 项目——华鑫中心
BELOW Itsuko Hasegawa Atelier exhibition area/project - Huaxin Center

172

上图	SANAA 建筑事务所漕河泾东区规划项目方案
TOP	SANAA Architectures' solution for the planning of east Caohejing
中图	GMP 建筑师事务所漕河泾东区规划项目方案
MIDDLE	GMP Architects' solution for the planning of east Caohejing
下图	大卫·奇普菲尔德建筑师事务所漕河泾东区规划项目方案
BOTTOM	David Chipperfield Architects' solution for the planning of east Caohejing

左图　智慧城市展展览现场
LEFT　Smart city exhibition site

成为新一代高新技术产业基地、现代服务业聚集地和国际一流水准的多功能综合性科技产业社区。12月08日开展的4号地块开发文献展，按照"4号地块"开发的时间顺序，选取了部分内容展览展出，并通过"前期策划""初步研究""国际竞赛""导则编著"四个展区内详实的图文资料、模型以及动画，向公众展示城市更新的先驱代表——华鑫股份在实践过程中不懈的努力和专业的精神。

装置艺术

韧山水

11月20日展出的《韧山水》是采用竹钢材料制作。线性杆件构成波浪起伏的两个空间，隆起的弧形高低错落，宛若山水，极具时代气息。杆件夹缝中隐藏了水雾喷淋，云雾缭绕的景象更强化了作品犹如城市山水盆景的视觉印象。它以柔软轻巧的姿态融入广场空间，巧妙地回应并契合了周围空间环境。建筑装置在城市公共空间中可以被理解成为一组开放式的景观。它既是艺术展品，同时也承载了容纳活动的实用功能。它激发着华鑫园区的活力，展现出新一代科技创意社区的开放精神。

华鑫天地"6+1"装置艺术展

以建筑师对城市空间的理解和各自的建构语汇设计一个或一组具有一定使用性的城市家具小品或装置。创作本身需回应漕河泾"新一代后工业社区"的概念以及"智慧城市"对未来社区所带来的影响和改变。

论坛及活动

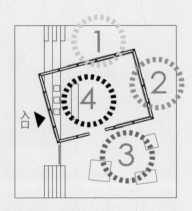

1——前期策划阶段展区
　　Early planning stage exhibition area
2——初步研究阶段展区
　　Preliminary study stage exhibition area
3——国际竞赛阶段展区
　　International bidding stage exhibition area
4——导则编著阶段展区
　　Words editing stage exhibition area

上图　4号地块开发文献展展区示意图
ABOVE　Plot No.4 development literature exhibition map

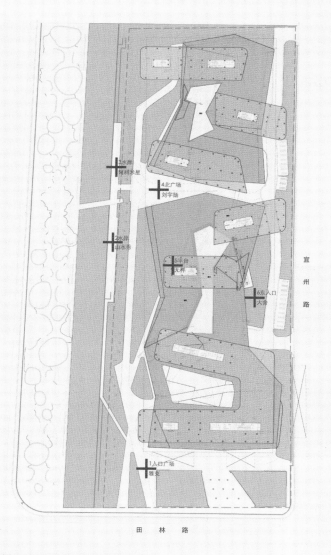

左图 6+1 装置艺术展平面
LEFT 6+1 installation art exhibition plan view

开幕论坛：自有其道的城市更新

《时代建筑》杂志运营总监、责任编辑、Let's Talk 创始人戴春主持"自有其道的城市更新"论坛，独立建筑师、Let's Talk 创始人俞挺、世界建筑副主编、清华大学建筑学院副教授周榕、贝诺中国首席代表庞嵚，中央美术学院建筑学院副教授、2014WAACA 中国建筑奖 - 社会公平奖优胜奖得主何崴，高目工作室主持建筑师、上海规委专家张佳晶分别发表了精彩的主题讲演，以各家之言，解析当前城市更新环境下的"自有其道"。

建筑师俞挺发表"从 Let's Talk 到 Let's Work"的主题讲演；世界建筑副主编周榕发表"硅基文明语境下的城市衰解与重构"的主题讲演；贝诺中国首席代表庞嵚发表"移植拯救城市"主题讲演；中央美术学院建筑学

1	3
2	4

1 明暗之间（无样建筑工作室）
   "Between Dark and Light" by Wuyang Architecture

2 竹屏（大舍建筑事务所）
   "Bamboo Screen" by Atelier Deshaus

3 气候之窗（法国雅克费尔叶建筑事务所）
   "Window of the Climate" by Jacques Ferrier Architectures

4 极度惊喜（周啸虎及其团队）
   "Surprise" by Zhou Xiaohu and his team

176

上图　4号地块开发文献展览现场
ABOVE　Plot No.4 development literature exhibition site

左图　智慧城市展展览现场
LEFT　Smart city exhibition site

右图　装置艺术作品《韧山水》
RIGHT　Flexible Landscape

上图　开幕论坛互动对谈：城市更新的多域视角
ABOVE Opening Forum: Urban Regeneration in Its Own Way

院副教授何崴发表"杂交生产，值得城市借鉴的乡村更新"主题讲演；高目工作室主持建筑师张佳晶发表"围城"主题讲演。

作为此次开幕论坛中的重要一环——Let's Talk 论坛，承办方秉承历届论坛的一贯品质，以"一个学术演讲 + 多位嘉宾对谈"的研究性专题研讨，来呈现前沿话题与观点的公开讨论。本次讨论围绕"城市更新"展开，有现今大环境的一番讨论，有北京、上海等各城市小环境的热议，有建筑观点的交流，有对未来的展望。

中期论坛：创建智慧城市
12 月 08 日，以"创建智慧城市"为主题的中期论坛在华鑫慧享中心举行。讲座由建筑师刘宇扬主持、智慧城市专家 Raoul Bunschoten 主讲。中期论坛开幕前，由策展人刘宇扬、马卫东导览参观了"智慧城市展"和"4 号地块开发文献展"。

当日下午 3 时，智慧城市讲座正式拉开帷幕，由刘宇扬建筑事务所创始人、主持建筑师香港大学建筑学院荣誉副教授、上海青浦区规划局顾问建筑师刘宇扬主持，CHORA 建筑与城市规划设计事务所创建者、负责人、德国柏林工大可持续城市研究中心主任 Raoul Bunschoten 教授主讲。

现场氛围严谨且轻松愉悦，Raoul 教授以智慧城市六讲为题，向大家介绍了其团队参与的智慧城市案例和智慧城市的概念，并郑重提出了气候问题。令人印象深刻的是，将智慧城市的深刻研究以骰子这种玩具形式呈现，使其智慧城市的研究探索中有了市民的参与，从一个看似只有专家才能懂得的理念，转化为人人可参与、触手可及的城市发展事业。

主题讲座结束后，以"智慧城市"为核心，举办了特邀嘉宾对谈。本次对谈呼应中期论坛的主题——创建智慧城市，列席的特邀嘉宾有华鑫股份总经理曹宇、万科集团总规划师付志强、文筑国际创始人马卫东、集合设计主持建筑师卜冰、阿科米星建筑设计事务所合伙创始人庄慎、同济城市规划设计研究院城市发展研究中心主创规划师苏运升、一宇景观设计建筑事务所主持景观建筑师林逸峰、都市工作群创始合伙人潘陶。

华鑫园区漫步系统
11月20日开始的"漫步"系统活动，结合艺术与科技的特色优势，以系统化设计为导向，综合解决信息传递、识别、辨别和形象传递等功能以帮助漫步在园区内的观众能够在最快的时间获得所需要的信息，加强观众的参与性和艺术体验，让观众在艺术氛围中感受漕河泾新兴技术开发区的魅力。

集思吧
集思吧长12米，宽5米，能够提供20～30人的活动空间。集思吧空间明亮，拥有挑战传统思维、激发想象和鼓励互动的基础设施，适宜举行各类小型会议、进行交流学习活动。极思吧提供了一个崭新的平台，将各路创新工作者汇聚在一起，自由交流，相互学习。

上海众创盛会——space+ 集市
结合创新创业分享团队、创业团队成果互动、微创业者市集，在公共空间中举办 space+ 集市，打造漕河泾的"创智天地"。

"书声"
当讲者诉说自己对书籍的思考和感悟时，对他而言就是一种深度的阅读，于听者是获取别人的思考与感悟，其实也是一种立体的阅读。"书声"是听书的道场，说书的擂台。

左图 列席"中期论坛——创建智慧城市"对谈环节的嘉宾（从左到右）：刘宇扬、卜冰、马卫东、付志强、曹宇、Raoul Bunschoten、庄慎、苏运升、林逸峰、潘陶

LEFT The attended guests to the dialogue session of the mid-term forum: The Creation of Smart City (from left): Liu Yuyang, Bu Bing, Ma Weidong, Fu Zhiqiang, Cao Yu, Raoul Bunschoten, Zhuang Shen, Su Yunsheng, Lin Yifeng, and Pan Tao.

"寻梦"摄影展及昆曲表演

展览及活动探讨了文化遗产在当代城市生活中的作用。进入了21世纪，在园子里唱曲听曲的机会已十分难得。但只要笛子吹响，檀板轻敲，曲人启喉开声之际，无论身处何地，无论多嘈杂的环境，昆曲都能奇妙地将人们带入到园林景致里头，让人体会到那一份渐已遗失的风雅。欣赏昆曲这样古老美好的艺术实际上并不需要太深入的了解，随着委婉的唱腔和精美的唱词，听者自然被引领进入梦境。

巴金摄影展

作品以上海的文化名人巴金的故居为线索，试图探讨名人、普通人以及时间对于城市建筑的影响。

左图　"寻梦"展户外昆曲表演
LEFT　"Dream Quest" outdoor Kun Opera performance
右图　"寻梦"摄影展现场
RIGHT　"Dream Quest" photography exhibition

上图　集思吧——虹梅庭分中心
ABOVE Jisi bar - Hongmeiting parallel session
下图　巴金旧居摄影展
BELOW Ba Jin's former residence photography exhibition

Theme of Case Exhibition

Through the onsite experience of Space Tour, visitors will re-understand the growing and development process of new communities, and, with people in those communities as the focus, through experience of the architectures in the block, re-understand the role that new technologies play in creating the community and inspiring new art forms.

The field work in the community is also carried out to identify cases for new community organization to be showcased more systematically, and to highlight the overall atmosphere of the new community with the exhibition on recent years' achievements in development and construction. Besides, the Space Tour route within the whole area is identified for visitors to explore the block. Architects and designers are invited to guide the tour. And the public space guiding system is also improved for the block. A combined team of architects and artists is invited to create customized key sites and spaces along the space tour route, and to form artworks and shows in different scenarios. This space work will act as a stage for the future public show center of the community, where public activities for the community will be organized on a regular base, thus forming the long-term mechanism of youth culture promotion where the exhibition case is located.

Exhibition Introduction

In the context of the transformation from the industry-driven economic model to a service-driven one, Huaxin Holdings, with five years' efforts, has achieved fruitful results based on a series of urban update practices in the Caohejing area. This exhibition as a part of the three exhibitions (i.e., the urban revitalization practice exhibition, the smart city exhibition, and Plot No.4 literature exhibition) of the Community and Technology living formats practice art exhibition in China Fortune Park for SUSAS, reflects plenty of attempts and ideas in the practice of urban update by Huaxin Holdings. It is not only the opening exhibition of the community and technology living formats practice art exhibition in Huaxin Park, but also the prologue of many urban update practices by Huaxin Holdings. Furthermore, the two art installation exhibitions - China Fortune Park Installation Art - Powerful Landscape Exhibition and the China Fortune World 6+1 Installation Art Exhibition also activate the space through the integration of art into spaces.

Urban Revitalization Practice Exhibition of China Fortune

The exhibition comprises three parts: Exhibition Hall of CHJ General Plan, four independent project exhibition halls and Exhibition Hall of International Competition Plans for Planning and Designing the CHJ East Zone (Plot IV).

The four independent exhibition halls present the public with pictures and articles as well as solid models from Itsuko Hasegawa Atelier, Green Pine Garden, Atelier Deshaus and Jacques Ferrier Architects in France. There, audiences can also appreciate competition plans from three world-top design offices: SANAA, David Chipperfield and GMP.

Smart City Exhibition

Divided into two exhibition halls, the Smart City Exhibition is curated by Professor Raoul Bunschoten, a famous smart city expert from Europe and TUB (Technische Universität Berlin)

Research Laboratory of Sustainable Cities under his lead. Exhibition 1 mainly presents "Six Lectures on Smart Cities"; Exhibition 2 discusses possibilities of developing Shanghai into a "Smart City" taking Berlin Smart City (Germany) as example.

Professor Raoul Bunschoten is the founder of CHORA in UK, a famous international expert in smart city. He is also a consultant in urban sustainable development and energy efficiency employed by Ministry of Construction in Germany and a foreign expert in National Development and Reform Commission of State Council of the People's Republic of China. Prof. Raoul Bunschoten and his team used to offer the smart city solution of the east of the Thames for London, UK, won the first prize of Berlin Tempelhof Airport Urban Modification Competition, and were invited to present smart city installation models for Shenzhen Biennial and Chengdu Biennial 2011 by Xiamen planning bureau.

Document Exhibition of Plot 4

Under the guidance of the national 12th five-year industry development plan, Plot No.4, located at the east area of Caohejing Development Zone, the southwest of downtown Shanghai, is striving to become a new-generation high-tech industrial base, a cluster of modern service industry, and a world class multifunctional and comprehensive scientific industrial community through its own industrial development, transformation and upgrade. Document Exhibition of Plot 4 was open to public on December 8, where part of developing achievements of Plot 4 are displayed in chronological order. Meanwhile, detailed articles and pictures in four exhibition halls of "Preliminary Planning", "Initial Study", "International Competition" and "Guide Preparation" show the public with the constant endeavors and professional spirits of Shanghai China Fortune Co., Ltd. (China Fortune) - a pioneer in urban regeneration.

Installation art

Flexible Landscape

Flexible Landscape presented on November 20 is a wooden bamboo structure. The two waving spaces made of linear beam elements deliver a sense of the times with landscape-like ridge and rugged arcs. The mist-shrouded scene created by the water mist sprays hidden in the gap of elements enhances the visual impression of urban landscape bonsai. Architecture installations in the urban public space can be regarded as a group of open landscapes. They represent both artworks and functions for real-life activities. This work gets involved in the square space and cleverly responds to surrounding environments in a soft and light profile, symbolling vitality of China Fortune Park and opening spirit of this new-generation scientific and creative community.

China Fortune Business Plaza "6+1" Appliance Art Exhibition

Architects and artists design a piece of or a collection of furniture item or appliance of certain value in use out of their conception on urban space and specialized vocabulary. The work itself embodies the concept of "New Generation Post-industrial Community" sweeping over Shanghai Caohejing Hi-Tech Park and reflects the impacts of "Smart City" on future communities.

Forums & Events

Opening Forum: Urban Regeneration in Its Own Way

A forum named "Urban Regeneration in Its Own Way" was held by Dai Chun, Chief of Operation and Editor in Charge of Time + Architecture, and founder of Let's Talk. Later, excellent and interesting keynote presentations were made by Yu Ting, Independent Architect and co-founder of Let's Talk; Zhou Rong, Associate Director of World Architecture and Associate Professor of School of Architecture, TSINGHUA University; Pang Qin, Chief Delegate of Benoy China; He Wei, Deputy Professor of School of CAFA, China Central Academy of Fine Arts, winner of 2014WAACA Social Equality Award; and Zhang Jiajing, Principal Architect of Atelier GOM Architecture and expert in Shanghai Planning Commission. They all shared their ideas about the theme of "Urban Regeneration - Just as It Needs to Be" in the context of existing urban regeneration environment.

Architect Yu Ting made a thematic speed named "From Let's Talk to Let's Work".

Zhou Rong, Associate Director of World Architecture addressed the speech of "Urban Degradation and Reconstruction under Silicon-based Civilization Context".

Pang Qin, Chief Delegate of Benoy China delivered a speech named "Survive Cities with Transplantation".

He Wei, Deputy Professor of China Central Academy of Fine Arts, gave a lecture of "Hybrid Production - What's Worth Learning from Rural Regeneration by Urban Regeneration"; and

Zhang Jiajing, Principal Architect of Atelier GOM Architecture, delivered a speech of "Fortress Besieged".

Let's Talk forum is an important part of this opening ceremony forum. The organizers have adhered to the consistent quality of past forums to organize public research project discussions in the pattern of "One academic speech + discussion of several guests" to present frontier topics and opinions. This discussion, surrounding "urban regeneration", covered discussion on current big environment, heated debates of small environments in cities like Beijing and Shanghai. Through this discussion, participants shared their architecture opinions and outlooks of the future.

Mid-term Forum: Creation of Smart Cities

On December 8, the mid-term forum themed "Building a Smart City" was held in China Fortune Huixiang Center.

The presenter was Architect Liu Yuyang and keynote speaker was Smart City expert Raoul Bunschoten.

Before its opening, curators Liu Yuyang and Ma Weidong visited the "Smart City Exhibition" and "Document Exhibition of Plot 4 Development".

At 3:00 p.m., Smart City Forum was unveiled to the public and was held by Liu Yuyang, founder of Atelier Liu Yuyang Architects, Principal Architect, Honorary Associate Professor of Faculty of Architecture, The University of Hong Kong, and Consulting Architect of Planning Bureau of Qingpu District. The keynote speaker was Professor Raoul Bunschoten, the founder

左页　爱运动的小伙伴（阿科米星建筑设计事务所）
OPPOSITE　"Sport Lovers" by Atelier Archmixing

and person in charge of Chora Architecture and Urbanism and Director of TUB (Technische Universität Berlin) Research Center of Sustainable Cities.

Professor Raoul started from six lectures on smart cities and introduced the concept of smart cities and related cases that his team had ever participated in a rigorous yet enjoyable atmosphere. In addition, he put forward climate problem. It is very impressive that the profound study of smart cities is embodied by dices, a game for every citizen to participate in the study. A concept which appears that only experts can understand is able to be transformed into an urban development cause open to everyone.

After thematic seminar, special guests were invited for discussion surrounding "Smart City". This discussion responded to the theme of this mid-term forum - Building a Smart City. Special guests present included Cao Yu, General Manager of China Fortune; Fu Zhiqiang, Chief Planner of VANKE; Ma Weidong, founder of CA GROUP; Bu Bing, Principal Architect of One Design; Zhuang Shen, co-creator of Atelier Archmixing; Su Yunsheng, Principal Planner of Urban Development Research Center of Shanghai Tongji Urban Planning & Design Institute; Lin Yifeng, Principal Landscape Architect of Yiyu Landscape Design Architectural Firm; and Pan Tao, founding partner of an urban working group.

"Wandering in Space" system in China Fortune Park

Oriented at systematic design, the "Wandering in Space" system activities initiated on November 20 combine the special characteristics of arts and technologies and provide a comprehensive solution for information transfer, identification, distinguishing and image transfer. There, audiences wandering in the park can obtain information in a shortest time period and are encouraged to participate in and experience arts and fell the charms of CHJ in art environment. This activity will last for the whole exhibition period.

Brainstorming Bar

Brainstorming Bar is 12 m long and 5 m wide and can accommodate about 20-30 persons. In this bright and well-equipped space, people are encouraged to challenge traditional thinking, inspire imagination and participate in interaction. It is a good space for holding small meetings of various types and for communication and learning. This bar provides a brand-new platform to gather innovation workers from all sectors to communicate with and learn from each other.

Shanghai Public Creation Celebration - space+ Fair

This is a public space for micro-entrepreneurs, where innovative and entrepreneur teams can share their achievements. It is to be created into a space of innovation and wisdom in CHJ.

Sound of Books

When the speaker tells about his thinking and feeling towards books, it is actually an in-depth reading for himself; the audience, who listen to such thinking and feeling told by others, are also enjoying 3D reading. Sound of books is where to listen and talk about books.

Dream Quest Photography Exhibition & Kun Opera Performance

At the exhibition and in related activities, the role of cultural heritage in the modern urban life is discussed. In the 21st century, you can hardly get the opportunity to sing or listen to

上图　space+ 集市
ABOVE Space+ fair

下图　主题活动现场
BELOW Thematic activity

traditional operas. But as long as the flute blows, the clapper flicks, and the performer begins to sing, no matter where you are, in a noisy or quiet environment, the Kun Opera will always take you into a fantastic garden scenery, to taste a gradually receding elegance. It doesn't need much understanding for you to enjoy Kun Opera, a wonderful ancient art. The graceful singing and refined libretto will naturally lead the audience into a dreamland.

Ba Jin Photography Exhibition

The creations take the Shanghai cultural celebrity Ba Jin's former residence as a clue, and try to discuss the impact of celebrities, ordinary people and time on the urban architecture.

策展人感想

刘宇扬

这次公共空间艺术季主题是城市更新,非常合乎时宜地回应中国城市化进程面临的全新挑战,这个挑战对与城市开放、设计、建设和管理者都是全新的研究课题。

这次的漕河泾案例展就是这样的一个实践导出研究的成果:关于漕河泾的前世今生,看华鑫置业,作为一家有高追求的企业、有高眼界的领导、有强大执行力的团队,带上一群给力的建筑师,如何把一个园区打造成了社区。

此次策展由马卫东、笔者和王家浩共同策划,主题是空间漫游。空间营造是建筑师擅长的本行,但我们希望把空间融入社区,从空间营造进化到社区营造,创造了新一代的社区概念:不仅居住的地方才是社区,工作、休闲、购物、文化的地方都可以是社区概念,这也是符合华鑫作为新一代智慧城市建设者的理念。

此次华鑫"社区与技术"生活业态实践案例展对"城市更新"主题进行了三个不同层面的展示与解读:开幕论坛"自有其道的城市更新"与"华鑫城市再生实践展"是对过往与当下经验的总结;中期论坛"创建智慧城市"邀请了欧洲著名可持续智慧城市专家 Raoul Bunschoten 教授,并结合四号地块展,表达了对智慧与未来城市的展望;闭幕论坛结合展览画册,A+U 华鑫专辑的出版以及华鑫天地"6+1"艺术装置展,则是"智慧城市"理念的实验与操作。

Note from the Curator

Liu Yuyang

"Urban Regeneration", the theme of this public space art season, responds to the brand-new challenges during the process of urbanization in China, which are new study topics for urban openness, design, construction and administrators. This CHJ case exhibition is a study-based practice achievement about the past and present of CHJ. From this exhibition, we learn how an industrial park is built into a community by a high-pursuit enterprise - China Fortune, with joint efforts of its forward-looking leadership and superior executive team, together with a group of outstanding architects. Ma Weidong, Wang Jiahao and I co-curate this exhibition with the theme of "Wandering in Space". Space construction is the architects' line of work. However, we hope to integrate space to communities and create a new generation of community concept from space construction to community building: a community is not only a place where we live but where work, leisure, shopping and culture are involved - this is also a concept that a generation of smart city constructor China Fortune will adhere to. This "Community & Technology" lifestyle practice case exhibition presents and interprets the "Urban Regeneration" from three different levels: The opening ceremony named "Urban Regeneration - Just as It Needs to Be" and the "China Fortune Urban Regeneration Practice Exhibition" are the summary of past and present experience; in the mid-term forum of "Building a Smart City", Professor Raoul Bunschoten, a famous sustainable and smart city expert from Europe is invited to express outlooks for wisdom and future city in combination of the Plot 4 Exhibition; and the closing forum, together with publication of exhibition picture album "A+U China Fortune Album" and "6+1" Installation Art Exhibition in China Fortune Business Plaza, are the experiment and practice of the "Smart City" concept.

策展人 CURATOR

马卫东 Ma Weidong
文筑国际创始人,上海市建筑学会理事,从事建筑设计、项目管理、建筑出版传媒,致力于中国建筑与世界建筑的互动与交流。
Professor of College of Architecture and Urban Planning, Tongji University; Master and PHD of Department of Architecture, Faculty of Engineering, The University of Tokyo; Founder of Director of CA GROUP; Director of The Architectural Society of Shanghai China.

刘宇扬 Liu Yuyang
刘宇扬建筑事务所主持建筑师,香港大学建筑学院副教授,上海青浦区规划局顾问建筑师。
Bachelor degree of Department of Urban Studies and Planning, University of California; Master of Architecture of Harvard University Founder/Principal Architect of Atelier Liu Yuyang Architects; Honorary Associate Professor of Faculty of Architecture, The University of Hong Kong; and Consulting Architect of Planning Bureau of Qingpu District.

王家浩 Wang Jiahao
建筑师、建筑批评及策展人、艺术家。2001 年成立加号(建筑 & 艺术)实验室,2006 年增设加号(城市 & 媒介)实验室。
Architect, Architecture Critic & Curator, Artist. He founded the + (Architecture & Arts) Lab in 2001, and + (City & Media) Lab in 2006.

主办单位 SPONSOR

徐汇区人民政府
People's Government of Xuhui District

承办单位 UNDERTAKER

徐汇区规划和土地管理局 | 上海华鑫股份有限公司
Xuhui District Planning and Land Administration Bureau | China Fortune Co. Ltd.

协办单位 SUPPORTER

虹梅庭公益服务中心
Hongmeiting Commonweal Service Center

地点 LOCATION

华鑫中心桂林路 406 号(宜山路交叉口)
China Fortune Center (No.406 Guilin Road)

文中图片均由徐汇区规划和土地管理局提供
All images in this case are provided by courtesy of Xuhui District Planning and Land Administration Bureau

互动水乡
朱家角·尚都里实践案例展
Interactive Rivertown
Site Project of Zhujiajiao·Shangduli

朱家角古镇位于上海市青浦区，属于典型的江南水乡风貌。它有着千年历史，历史上曾以布业著称江南，以米业兴旺千家，是著名的富饶水乡。时代交替，褪去繁华，如今的朱家角是上海著名的旅游风景区，有上海威尼斯之称。

尚都里，位于古镇三板块（古镇保护区、老镇协调区和新镇控制区）的交汇处，开发前曾是油脂厂和民居。项目立项之初，政府和开发单位就"如何改造或重新开发，以复兴繁华江南"反复论证，最终确定了"时尚江南"这个定位。

江南，从古至今便是全国的时尚中心，文人墨客聚集之地。在今天，它的时尚韵味也不应没落。尚都里，以现代设计打造一个江南建筑群，并在建筑群内入驻时尚商业，以"繁华"复兴这座正渐渐褪去光彩的江南巨镇。

建筑上，古镇保护区域是十分典型的明清江南水乡建筑群，有着错综交汇的市井深巷、几近随机的大小院落、依偎水边的楼阁亭台。为了尽可能地保留这传统风貌，同时表现现代化发展形成的时尚生活方式，开发单位大胆对项目场地范围内功能重新设定，并在建筑风格上采取了化整为零的做法，通过单纯的建筑用材配以多变的制造工艺，创造出古镇朴实玲珑的建筑风格以及纷繁多变的城市肌理。

项目分解出5块区域，由5位具有国际背景的华人建筑大师分别打造。5个区域既要互相照应形成统一的整体氛围，又要在每个策略上注重地方材料的运用和地方工艺的延伸，突出其个性。这不仅仅是一次商业的开发，更是一次江南水乡的复兴运动。

由原先的工业文明和农属产业遗留下来的建筑群，具有一定的历史文

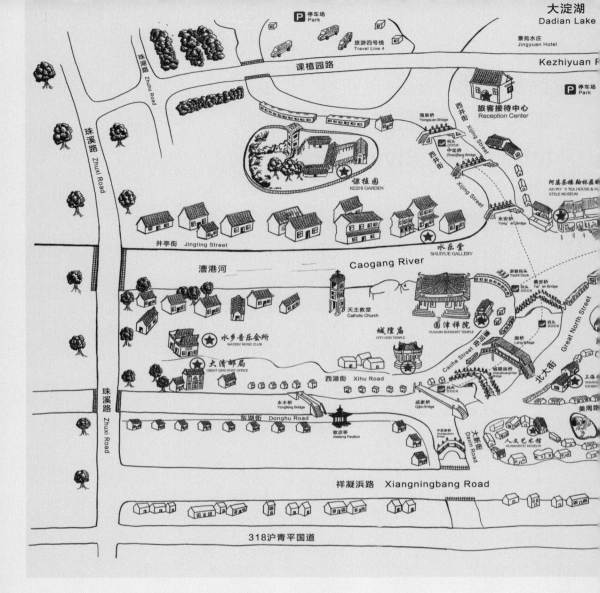

化意义，所以在改造时，保留了一部分原有的建筑物来点睛。如经过改造的油灌成为现代建筑的一部分，美观又具有实用意义的连廊，原木搭建而成的榫卯结构等，这些从江南母体文化中幻化出来的点睛之笔，将在未来作为建筑研究的范本，也是这座城市复兴的新起点。

Zhujiajiao, located in Qingpu District, Shanghai, features a waterside landscape typical in the south of the Yangtze River. This millennium-old watery city, once a textile hub and rice capital known for its amazing wealth and large population, is a famous attraction in Shanghai, known as the Venice of Shanghai.

Shangdu Lane, located in the cross point of Ancient Town Preservation Zone, Old Town Co-ordination Zone and New Town Control Zone, was occupied by oil mill and residence zone.

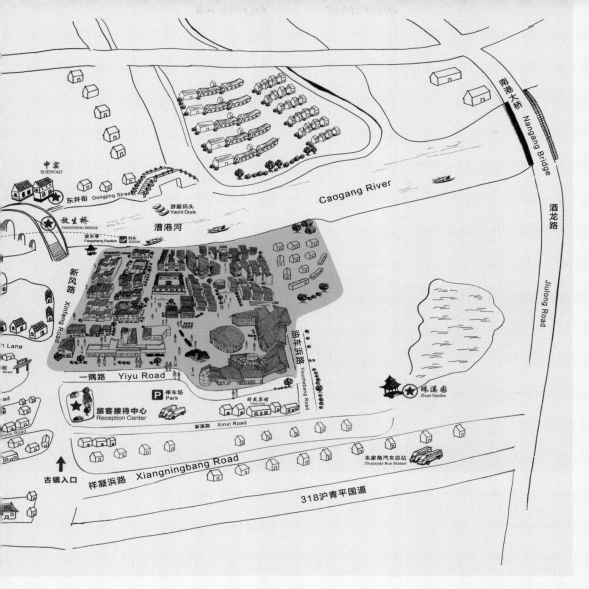

At the beginning of the project, the government and developers demonstrated repeatedly on how to transform or re-develop this lot to revive the South of the Yangtze River. They ended up with a fashion-oriented position.

The Jiangnan (means the regions south of the Yangtze River), an all-time fashion center and favorite haunt for writers and poets of China, should leave no fashion charm fading away. Shangdu Lane hosts a modern-design commercial architectural complex to revive this waning capital with updated glamour.

The ancient town preservation zone, architecturally, is characterized by the signature style in Ming (1368–1644) and Qing dynasties (1616–1912) in waterside towns in Jiangnan, a proud owner of crisscross alleys &lanes, a hodgepodge of yards and waterside pavilions. To conserve the traditional landscape and showcase the modern fashion life style, the

上图 尚都里项目位于朱家角古镇"古镇保护区、老镇协调区、新镇控制区"三版块交汇处

ABOVE Shangdu Lane Project is located in the cross point of Ancient Town Preservation Zone, Old Town Coordination Zone and New Town Control Zone

developers redesigned the functions in the project site and create aquaintand sophisticatedbuilding style and rich urban textures chunk by chunk with simple building materials and abundant techniques.

The project, in five parts, will be designed by five Chinese international architects respectively to integrate them into one entity with harmonious atmosphere and strategically highlight their unique identities by local materials and techniques. This is more a revival of the waterside town in Jiangnan than a commercial development project.

Some of the historically significant remaining of ancient industries and agriculture are conserved in transformation as the focal point of the project. For example, the transformed oil tanks become a part of modern buildings, so are the beautiful, useful corridors and tenon-and-mortise wooden works. Those highlights derived from the native culture in Jiangnan, will serve as role models for architectural researches and the starting points for city reviving.

展览主旨

"尚都里·互动水乡实践案例展"以朱家角传统文化历史风貌与尚都里新江南水乡当代城镇的发展融合为依托，强调公共空间发展和人文传

上图 从放生桥望去，都里在左，古镇在右
ABOVE A view from Fangsheng Bridge: Shangdu Lane on the left, ancient town on the right

统发展的时间线，使它们直面并相融。展览通过现代公共空间的艺术语言，寻找有机融合的切入点，创造城市与乡村坦然面对的机遇。案例展通过不同展区和展示内容，传递人与物之间更直接的关系，突出江南水乡古镇传统名俗与当代生活的现状。

展览介绍

案例展通过 1 个室内主展馆、3 个室外街区展馆、多个活动的形式，打造一个全新的拥有丰富现代互动艺术的水乡空间。来自不同领域的艺术家通过自己对水乡的独特理解，塑造出姿态各异的艺术作品，并通过作品与人的互动给观众带来关于水乡的全新体验。

室内主展馆：龙珠艺术中心

由 Philip Beesley（滑铁卢大学建筑学院教授，建筑和数字媒体艺术的实践者）创作的声光互动作品"春龙条"，以遍布雕塑的传感器感应游客的存在，设置精致柔和的光波、低语的声音、如皮影般的戏剧性效果，让观众亲密体验个人、艺术品和空间的融合。

艺术中心同时展出现代3D打印作品,分别由Steven Ma、Rem D Koolhaas和DADA数字专业委员会带来。

三个室外街区展馆:现代水乡街巷 + 手工艺复兴街区 + 尚窗艺术街区

现代水乡街巷:以"现代互动艺术"为主题,体现现代水乡空间与人的互动性,新的水乡空间鲜活地存在于市民的生活之中。街巷展示了美国伦斯勒理工建筑学院院长Evan Douglis的"声·云"艺术装置、墨尔本皇家理工大学教授Roland Snooks的"集群建造"作品、HHDFUN事务所合伙人王振飞和王鹿鸣的"城市家具"等。

"声·云"艺术装置,作品将声音作为一种主要变革体验的媒介手段来给人们提供一种完全身临其境的感受,回味我们生活中那些对于声音的记忆,这些记忆保存着过去、现在和未来对于我们的重要性。"集群建造",通过协作机器人的交互协调制作的户外雕塑装置。"城市家具",以装置的方式探讨人与城市空间的关系,座椅单元可以进行复制、镜像或旋转,进而拓扑组合产生无限可能,以适应不同的城市空间需要。

在现代互动艺术作品展现魅力的同时,国内外壁画艺术家以水乡特有的大幅白墙为基调,创作现代水乡墙绘作品,并以系列形式展出。参与的艺术家包括:原田透(日本壁画艺术家)、icy and sot(美国艺术家)、李秉罡(美籍壁画艺术家)、文刀绘(本名:刘冀湘,中国壁画艺术家)等。

左图 室内主展馆"龙珠艺术中心举办开幕式"
LEFT Main indoor exhibition hall: Open Ceremony of Longzhu Art Center

右图 3D打印作品展
RIGHT 3D printed work exhibition

本页 Philip Beesley 声光互动作品"春龙条"
THIS PAGE Philip Beesley - sound-light interactive artwork "Spring Dragon"

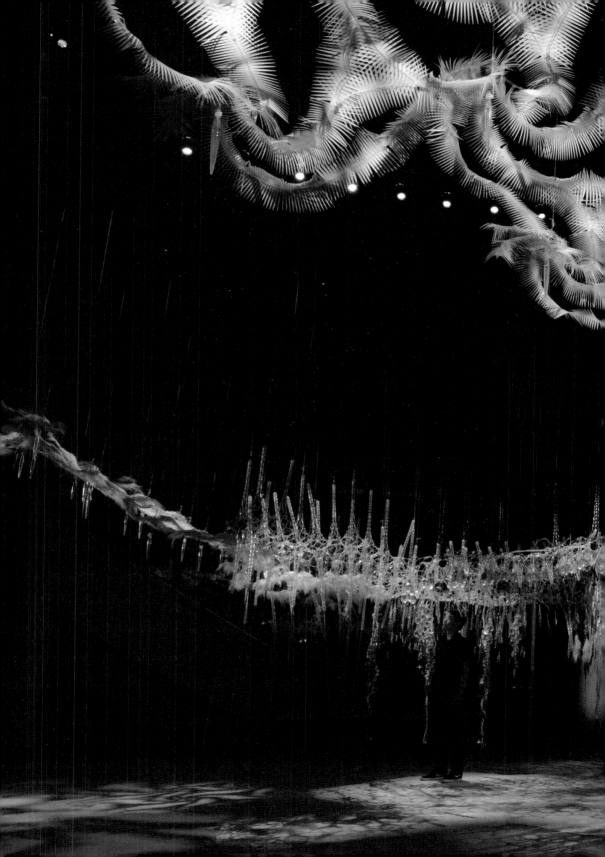

1	4
2	5
3	6

1　Roland Snooks 作品"集群建造"
　　Roland Snooks :"Cluster"

2　王振飞王鹿鸣作品"城市家具"
　　Wang Zhenfei& Wang Luming :"City Furniture"

3　时尚江南，水乡街巷
　　Fashionable Jiangnan Region with traditional watersidevillages and streets

4　薛林纳超现实主义作品"不朽的达利""毕加索")
Surrealistic artwork of XueLinna- "Immortal Dali" and "Picasso"

5　壁画艺术家原田透、Icy and Sot、李秉罡、绘墙作品
Works of frescopainters including Toru Harada, Icy and Sot, Li Binggang and Wen Daohui

6　Evan Douglis 声·云 互动艺术装置
Evan Douglis - sound-cloud artistic facility

上图　手工艺复兴街区
ABOVE Crafts revival block

著名雕塑家、画家和环境艺术家薛林纳在街巷中展示了"不朽的达利"和"毕加索"两个作品，通过超现实主义的个人风格创作，展现艺术大师对于水乡所产生的凝视和畅想。

手工艺复兴街区
8家手工艺单位参与，包括陶艺、木工艺、手作巧克力、手作烧饼等传统和创新内容。传统的手工技艺，创新的人文氛围，以极具创意的现代感推动传统手工艺的复兴，打造"传承、创新、复兴"的新人文手工艺氛围，以此唤起传统手工艺复兴之路。

尚窗艺术街区：尚都里建成以来，多位艺术家和设计师在此举办活动和个展，此次"尚窗艺术展"向游客展示多位艺术家和设计师的风采，展览于街区橱窗展示。

多个活动：开幕式论坛、海派旗袍大赛
案例展共举办了1次开幕式+论坛活动、3次朱家角海派旗袍大赛活动。9月30日，开幕式论坛邀请当年亲身参与其间的建筑和规划界专家再次聚首朱家角，张永和、登琨艳、马清运、柳亦春几位建筑大师出席。他们当年怀抱着浓厚的江南情结，秉承着开放的眼光，孕育出了"似曾相识，又有不同"的当代江南水乡。毫无疑问，建筑与技术从宏观层面影响着城乡的交替与更新；如今，他们回顾十年，以当代城市发展变革思路和古镇的新旧平衡为议题，重新审视尚都里项目对激发古镇活力所带来的影响。最重要的讨论在于：如何在日新月异的社会中，唤起人们对自我意识的觉知，活得更加糅合，自省，具有当代气度。

复兴水乡空间的同时，更要复兴水乡文化，加深居民对传统文化的情感，此次展览同期推出"朱家角海派旗袍大赛"，号召全市女性参与。10月25日，朱家角海派旗袍大赛举办初赛，约200位水乡淑媛参与；11月

31 日，举办复赛，共 50 位佳丽入围；11 月 07 日，举办决赛，决出冠亚季军和才艺奖，水乡淑媛尽展风采。

本页 观展人群
THIS PAGE Visitors and audiences

Theme of Case Exhibition

An Interactive Waterside Wonder—Layout of Zhujiajiao Shangdu Lane, based on the development and integration of the cultural and history legacy in Zhujiajiao and the modern waterside towns in Shangdu Lane, underlines the time frame for development on public spaces and cultural traditions and provides an opportunity of urban-rural outlook. This layout depicts thehuman-object relationships and reflects the folk customs and contemporary lives in a time-honored waterside town in Jiangnan.

Exhibition Introduction

Thislayout, with 1 indoor exhibition hall, 3 outdoor exhibition halls and various events, creates a waterside space rich in modern interactive art factors. Diverse artists create an array of works to expose audiences to new interactive experience in this waterside town.

本页　互动水乡风采
THIS PAGE　Interactive Watery region

Main Indoor exhibition hall: Longzhu Art Center

Spring Dragon Twigs (Chun Long Tiao), an acousto-optic interactive artwork created by Philip Beesley (Professor of School of Architecture, University of Waterloo) detects the audiences with its detectors overall its surface to present delicate and gentle light, whispers and shadow puppet alike dramatic effect to bring audience into an immersive and integrated art space.

The show center also presented modern 3D printed artworks created by Steven Ma, Rem D Koolhaas and Digital Architecture Design Association (DADA).

3 outdoor exhibition halls: Modern Waterside Lanes + Traditional Handicraft Plaza+ Shangchuang Art Street

Modern Waterside Lanes: This is a refreshing, interactive and modern waterside space themed as Modern Reactive Art and integrated into our daily lives. The spot displays "Sound · Cloud" art installation created by Evan Douglis (Director of Rensselaer Polytechnic Institute - School

本页　互动水乡风采
THIS PAGE　Interactive Watery region

上图　艺术家们正在制作《春龙条》
ABOVE The artists are making Spring Dragon

of Architecture), "Aggregate Construction" works created by Roland Snooks (Professor of Royal Melbourne Institute of Technology), "Urban Furniture" created by Wang Zhenfei and Wang Luming, HHDFUN partners and more.

"Sound · Cloud" is an artistic installation that transformatively provides the audience an immersive experience with sound and arouses their memories of sounds, which embody the meaning of past, now and future. "Aggregate Construction" is an outdoor sculpture installation constructed by interactive and coordinated robots. "Urban Furniture" casts light on the relationship of human and urban space with duplicated, mirrored or rotated chair units and infinite topological-combing potentials to meet urban space demands.

In addition to those modern interactive artworks, fresco artists have created serial modern wall paintings of waterside landscape on the signature white walls in the South of the Yangtze River. These artists include Toru Harada (Japanese fresco artist), icy and sot (American artist), Bing Lee (Chinese American fresco artist) and Wen Daohui (Autonym: Liu Jixiang, Chinese fresco artist).

Xue Linna, a famous sculptor, painter and environment artists, has presented his works "Eternal Dali" and "Picasso" to show the gazing and imagination over this waterside town of art masters with his surrealistic style.

Handicraft Rejuvenation Street/Block

8 handicraft workshops have settled down in the street, including traditional and innovative workshops such as a pottery workshop, a carpenter workshop, a DIY chocolate shop, and a Chinese baked roll stall. The combination of traditional handicraft art and innovative cultural atmosphere boosts the rejuvenation of traditional handicraft art in a super creative way and forms the new cultural handicraft atmosphere featuring "inheritance, creation and renaissance",so as to pave the road towards the rejuvenation of traditional handicraft art.

Shangchuang Art Street: since its completion, Shangduli has cradled events and solo shows of many artists and designers. For example, the current "Shangchuang Art Show" is a great opportunity when visitors can enjoy marvelous works of many artists and designers kept in the showcases along the street.

Events: opening ceremony & forum, Shanghai style cheongsam competition

The project consisted of 1 opening ceremony + forum, and 3 rounds of Shanghai style cheongsam competition. At the opening ceremony and forum held on Sept. 30, architects and planners who had developed Zhujiajiao got together at Zhujiajiao. Great architects including Zhang Yonghe, Deng Kunyan, Ma Qingyun and Liu Yichun also took part in the event. Driven by their obsession with Jiangnan Region (namely the regions south of the Yangtze River), they fostered a familiar but slightly different modern Jiangnan with their broad visions 10 years ago. Undoubtedly, the modern architecture and technology affected the alteration and upgrade of cities and towns in a macroscopic way. Now, these masters reviewed their achievement 10 years ago and started to analyze how Shangduli stimulated the old town from the aspects of the development and reformation of modern cities and the balance of the past and the present. What's more, these masters wanted to find a much more important answer: at this volatile era, how to evoke people's self-awareness and help them to live a modern life featuring deeper blending and self-examination.

While we regenerated the waterside town, it was more important to regenerate its culture and deepen people's love for traditional culture. Therefore, during the exhibition, "Zhujiajiao Shanghai Style Cheongsam Competition" was held and all female Shanghainese were encouraged to participate in this event. On Oct. 25, about 200 young ladies participated in the preliminary contest of the competition; then, on Nov. 31, 50 beauties took part in the second round; and on Nov. 7, competitors showed their excellence and the winners of the top three awards and the talent awards were selected.

左图　Roland Snooks《集群建造》
LEFT　Roland Snooks, "Cluster"

中图及右图　Evan Douglis《声·云》
MIDDLE & RIGHT　Evan Douglis, "sound-cloud"

策展人感想

袁烽

我们策展的主旨，是通过引入互动的方式，通过现代艺术与江南水乡的跨界碰撞，打破传统水乡现存空间的属性与维度，为千年古镇朱家角植入全新的意念与能量。让朱家角水乡在吴侬软语的温婉之中，迸发出现代、个性的新声音，展现其在发展过程中对水乡原真性的巧妙保留以及对当地艺术及人文的工艺复兴作出的贡献。这种互动是当代与历史，现在与未来的对话，是传统空间与当下实践的对话。尚都里将都市人的生产和生活方式融入江南水乡的文化价值之中，关注人与水乡环境，人与城市空间的相互关系，营造出新旧交融、有机生长、和谐共存的诗意古镇。此外，公共艺术的介入，改变了水乡的生活品质和状态，用当代的方式呈现出未来水乡的可能性。

薛鸣华

曾是一片厂房、仓库的残破之地朱家角·尚都里，四位建筑大师透过江南水乡情怀，秉承国际化开放的视野，孕育出当代的都市水乡。作为2015上海城市空间艺术季实践案例展之一的"互动水乡·十年回望"相聚朱家角·尚都里，共叙漕港河畔的十年历程。

十年前在此讨论传承与创造，十年后的今天，再次感受千年古镇的传统和当代的精致。工艺复兴的朱家角·尚都里以建筑、雕塑作品、绘画、影像等多种方式展现当代的都市水乡，用世界的语言诠释古镇的文化与特色。

走在古老放生桥畔，听潺潺漕港河水，看当代艺术作品，感场所更替变迁，回味无穷。

Note from the Curator

Yuan Feng

When we prepared the event, we intended to introduce interactive measures and combine modern art with traditional style of Jiangnan Region, so as to break the limit of traditional waterside town in terms of its property and dimension and inject new concept and energy into the artery of millennial Zhujiajiao. By doing so, we wanted to add new notes of modernism and personality to the sweet Shanghainese dialect at Zhujiajiao, to reveal the true nature of this town reserved smartly during its development and to show the contribution of their design to the rejuvenation of local art and culture. What was embedded in this interaction was a dialog between the past and the present, the present and the future, as well as traditional space and modern practice.Shangduli blended the modern production and life of urbanites with the traditional culture and value of Jiangnan Region. Valuing the relationship between people and the waterside town, between people and urban space, we had built a poetic town where the past and the present mingled and developed in an organic and harmonious way. Besides, the introduction of public art had changed the quality and status of life in this waterside town and exhibited in a modern way what this town might be in the future.

Xue Minghua

Shanghai used to be a land of dilapidated plants and warehouses. However, under the hands of four great architects who obsessed about Jiangnan style and boasted international vision, this deserted land became a modern waterside metropolis. As a site project of Shanghai Urban Space Art Season 2015, "Interact with the Waterside Town & Review the Decade-old Achievement" was an opportunity when we met here at Shangduli in Zhujiajiao, and reviewed the past decade witnessed by the Caogang River.Ten years ago, we had talked about inheritance and creation. Ten years later, we again sensed the traditional features and modern delicacy of this millennial town. Shangduli at this era of handicraft rejuvenation was demonstrating the elegance of this modern waterside metropolis with buildings, sculptures, paintings and video clips. It was interpreting the culture and feature of the old town in languages understood by the whole world.When you walked by the old Fangsheng Bridge and captured the babbling Caogang River with your ears and modern art with your eyes, you would feel the change of this town and its endless aftertastes.

策展人 CURATOR

袁烽 Yuan Feng
同济大学建筑与城市规划学院副教授
国家教育部高密度人居环境实验室"数字设计研究中心"负责人
上海创盟国际建筑设计有限公司创始合伙人
中国建筑学会建筑师分会第六届理事
中国建筑学会建筑师分会数字设计专业委员会联合发起人
Associate Professor of College of Architecture and Urban Planning of Tongji University
Director of "Research Center of Digital Design" of High Density Residential Environment Lab of MOE
Co-founder of ARCHI-UNION
Member of the 6th Institute of Chinese Architects, ASC
Co-initiator of Digital Design Committee of Institute of Chinese Architects, ASC

薛鸣华 Xue Minghua
同济大学城规学院景观园林硕士 MLA
上海安墨吉建筑规划设计有限公司主建建筑师
MLA of College of Architecture and Urban Planning, Tongji University
Principal Architect of AMJ Architecture & Urbanism

地点 LOCATION

上海市青浦区朱家角新风路 240 弄尚都里内

Shangduli, Alley 240, Xinfeng Road, Zhujiajiao, Qingpu District, Shanghai

重新装载
浦江东岸老白渡码头
城市更新实践案例展
Reloading
Urban Renewal in Practice of Laobaidu Wharf, East Bank of Huangpu River

浦东陆家嘴老白渡码头曾经是上海市重要的煤炭卸载码头。今天它已经完成了过去的历史使命，而等待在城市更新中扮演新的角色。展览主题"重新装载 (Reloading)"，意为正在重新装载，具有丰富的含义。老白渡码头不仅是货物卸载，也是煤炭装载上车并送入市区的场地。码头通过卸载和装载的生产行动，给城市提供能量。老白渡码头装载煤炭的历史已经终结，现在它准备重新开始装载新的内容，向城市输送文化艺术的新能量。这正体现了城市更新的重要意义：把具有历史价值的城市基础设施转变成城市公共文化艺术空间，使城市不断激发出新的生命力。

Laobaidu Dock at Lujiazui, Pudong District was the most important coal port of Shanghai. Having completed its mission as a coal port, the dock is now waiting to make new contribution to Shanghai during its upgrade. The theme of the site project is "Reloading". The word boasts rich connotations, meaning to reload the dock with new content instead of coal. Laobaidu Dock was more than a port for unloading goods. It was also the place where coal was loaded to vehicles to the downtown area. In other words, the dock powered the city with loading and unloading. However, today, the dock is freed from this old mission. It is going to shoulder new content, namely, feeding the city with cultural and artistic energy. This change reflects the crucial meaning of urban upgrade: constantly stimulate the vitality of a city by transforming historic infrastructure of the city into cultural and artistic space open to the public.

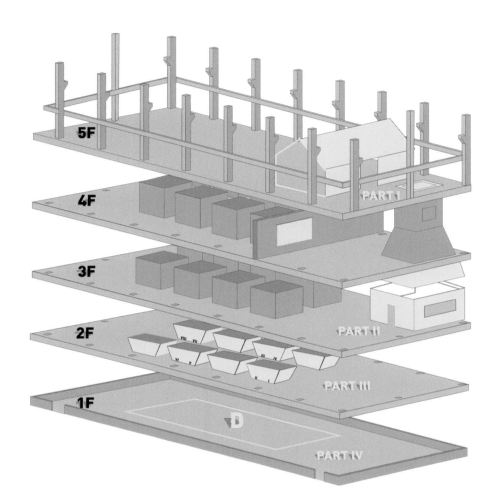

PART I 预备
PRELOAD

PART II 转换
TRANSITION

PART III 激活
ACTIVATION

PART IV 再生
REGENERATION

上图　开幕式现场
ABOVE Opening Ceremony

下图　研讨会现场
BELOW Seminar

展览主旨
....

老白渡煤仓及其廊道改造实践案例展将是一个动态的过程。在本次艺术季的过程中，原有的煤仓作为展览的现场，既包括老白渡煤仓和廊道在未来的实际改造方案，也包括配合本次艺术季展览主题"城市更新"的相关案例展。

展览介绍
....

本展览共有四个单元：分别为"预备"，"转换"，"激活"和"再生"。其中"预备"单元将选取、展示一批优秀的国内外工业历史建筑的改造案例，通过影像与文本文献，呈现其改造的理念与样态；"转换"单元则将专门呈现老白渡煤仓和廊道未来的改造方案；而"激活"单元，通过委托一群来自现代舞、声音、影像和多媒体等背景的艺术家，基于八个本地案例空间（现厂、韩天衡美术馆、雅昌艺术中心、上海电子工业学院、五维创意园 J-office、西岸艺术中心、外马路 1178 号创业办公、老白渡码头）各自的情况，联合创作完成一组影像作品，并以多媒体影像装置的方式呈现，以探索身体、影像和空间的关系；"再生"单元是把工业废墟转化为注入艺术能量的活动广场，利用声音和灯光装置作品进行特别设计。

本展从建筑空间出发，集合当代艺术的多种形式，以多元的方式，提供丰富的感知，进一步阐释现代人与现代空间和城市生活的动态关系，打造一个全新概念的建筑案例展。

第一部分 预备
本单元是整个展览的序厅，提出了展览所要讨论的观点及其语境，即：在城市更新建设的过程中，如何处理过去留下的工业建筑是一个必须面对的问题，而工业建筑也有超越装饰性之外的美学和价值。展览选取了一批优秀国内外工业历史建筑的改造案例，通过影像及文本文献，呈现其改造理念与样态。

左图　Part 1 预备 展厅层（陈颢摄影）
LEFT　Part 1 Preload exhibition hall (Photo: Chen Hao)

216

上图　Part 1 预备 展厅室内（陈颢摄影）
ABOVE Interior of Part 1 Preload exhibition hall (Photo: Chen Hao)

下图　Part 1 预备 展厅层（苏圣亮摄影）
BELOW Part 1 Preload exhibition hall (Photo: Shengliang)

左页 Part 2 转换 展厅内部（陈颢摄影）　　本页 Part 3 激活 展厅层（苏圣亮摄影）
OPPOSITE Interior of Part 2 Transition exhibition hall (Photo: Chen Hao)　　**THIS PAGE** Part 3 Activation exhibition hall (Photo: Shengliang)

Part 3 激活 展厅层 投影出的展览名称
Projected exhibition name in Part 3

上图　Part 4 再生展层（陈颢摄影）
ABOVE Part4 Regeneration exhibition hall (Photo: Chen Hao)

第二部分 转换

老白渡煤仓及廊道日后将会改造成一个公共文化艺术空间，本单元集中展示其未来改造方案。

第三部分 激活

本单元展示了八个本土的工业建筑改造案例，并由参展建筑师提供相关建筑模型。同时，还特别呈现了为本展委托创作的《呼唤》系列，它包括八部短片及一组特殊的影像声音装置。整组作品是基于这八个本土的改造案例，邀请艺术家从身体、空间和影像的角度，探索它们之间的关系，并通过合作作品予以回应。

《呼唤》八部系列短片 I 号至 VIII 号，分别在八个案例空间中进行表演和拍摄，并在每个案例及其内部不同表演空间内进行录音。试图从影像、身体和声音的维度，表现各空间内部的属性，以及各案例自身的特质。

上图 《呼唤》系列影像图
ABOVE Image map of "Evocation"

《呼唤·特别版》是一组结合老白渡煤仓特殊空间所设计的影像声音装置。在煤漏斗形成的特殊通道中，同步播放《呼唤 I-VIII》八个影片。现场使用了八个音箱，分别播放构成主题音乐的八个不同乐器 / 声音的音轨，它们既与八个案例分别呼应，又在现场合为一曲交响。

第四部分 再生

本单元展示了一组声音装置作品。不同年代交通工具的声音从外侧的四只煤漏斗里卸落而下，带来历史与当下的隐喻。在中心区域，超音波技术射出纯抽象的音束，使人感到听觉上的紧张和身体上的压迫。而当观众在展厅中移动时，这两种具象和抽象声音的边界则渐渐模糊。

Theme of Case Exhibition

The reconstruction site project of Laobaidu Coal Bunker and its corridors was dynamic. During the art season, the original bunker became the exhibition pavilion where visitors could find the plan for the future reconstruction of the bunker and its corridors, and the site projects related to the theme of the art season, i.e., "urban upgrade".

Exhibition Introduction

PART 1 Preload

Set in the lobby of the space, this section serves as an overview showing the core and the background of the exhibition, that is, the architect has to deal with historical industrial buildings during urban renewal to show their aesthetic values that is beyond being pure decorative. A group of outstanding cases of industrial architecture renewal from home and abroad are selected and displayed in both visual and textual archives to show the diverse approaches and results.

PART 2 Transition

As Laobaidu coal bunker and its gallery bridge will be turned into a public art space, this section gives a full display of the renovation plan.

PART 3 Activation

This section shows eight local cases of industrial architecture renewal with the models from the participating architects. Meanwhile, it also presents the work "Evocation", a specially commission for this exhibition, including eight serial short films and a set of image-sound installation. The whole idea of Evocation series is to invite artists to collaborate and to respond to the eight local renewal cases from the perspectives body, space and image, and the relationships between them.

Evocation serial videos I to VIII, are performed and shot in the eight venues of the selected cases, and the theme music has been played and recorded in every shooting site of every venue. By doing so, thevideos attempt to describe the inner diversityand physical features of each case fromvisual, physical and perspectives. EVOCATION · SPECIAL EDITION is a set ofvideo-sound installation designed for theunique corridor space within the Laobaiducoal bunker. Eight videos, Evocation I – VIII will be screened on original surfaces of thebuilding structure, in manner of synchronizeddisplaying. There are eight speakersplaying the eight instruments/sounds fromthe theme music which not only respond tothe eight cases respectively, but perform anlive orchestra.

Part 4 Regeneration

This section displays a set of sound installations. In the outer part, sounds of different transportations,standing for old and modern times, drop from the ceiling. While in the middle part, pure abstract sound delivered by ultrasonic wave lead tensions to the audience both acoustically and physically. When the audience move around the space, the line between the figurative and abstract sounds becomes blurred.

策展人感想

冯路

从工作室的落地窗看出去,恒隆广场和周边的高层建筑共同构成了一幅当代大都市的典型而优美的图像。然而,当我走在它们脚下,却只能沿着南京西路匆匆前行。虽然沿街的橱窗乃至街道景观都和建筑一样精致美观,但是,人的身体与其之间却像有着遥远的距离。在这精致如画的城市空间里,我们都是游客。这种视觉和身体感知上截然不同的对比,恰恰就是最近十余年间,上海新建的诸多城市空间的特征。城市空间景观化不仅可以看到视觉快速消费文化的影响,也是空间生产缺乏自主性的表现。上海的城市更新应该鼓励一种更加有活力的空间生产机制,它能够激发多样的空间文化而抵抗单一的精致主义,它能够唤起身体空间的主体性而摆脱视觉化的图像控制,它能够促成自我有机生长的上海而告别被他者欲望所投射的东方巴黎。

Note from the Curator

Feng Lu

From the scene of Plaza 66 and surrounding high-rises out of the French windows of my studio, I can see the unique features and elegance of a modern metropolis. However, when I walk on the road where these buildings stand, I can only hurry along the West Nanjing Road. Although the roadside windows and street views are delicate, these facilities seem to be far away from people' bodies. In this picturesque city, we are all passers. The great difference between what people see with their eyes and what they feel with their bodies is the typical feature of the vast urban space completed in the recent ten years in Shanghai. From this urban landscape, we can feel the influence of fast-food-like visual consumption, and the shortage of autonomy of space production. To promote urban upgrade of Shanghai, we need to encourage space production mechanisms that are more energetic and that can stimulate the development of diverse space culture, instead of the monotonous pursuit of delicacy, that can yield architectures which can be felt by people's bodies instead of just being enjoyable to people's eyes, and that can promote organic self-development of Shanghai, instead of forcing Shanghai to submit to others' lust and become Paris of the East.

策展人 CURATOR

冯路 Feng Lu
英国谢菲尔德大学建筑学博士,无样建筑工作室主持建筑师,南京大学和上海交通大学客座导师。
Doctor of Architecture of The University of Sheffield (UK), Chief Architect of Wuyang Architecture, and a guest mentor of Nanjing University and Shanghai Jiaotong University.

柳亦春 Liu Yichun
大舍建筑设计事务所创始合伙人,国家一级建筑师,同济大学建筑城市规划学院和东南大学建筑学院客座教授。
Co-founder of Atelier Deshaus, First Class Architect of China, and a guest professor of CAUP Tongji University and School of Architecture of Southeast University.

颜晓东 Yan Xiaodong
独立策展人、制作人。
An independent curator, and a producer.

合作艺术家 ARTIST PARTNERS

殷漪(作曲、声音装置)Yin Yi (Composition & Sound Installation)
实验音乐、声音艺术家,BM SPACE 联合创始人。
Owning to his various work and practice, Yin Yi is now a key player in the field of experimental music and sound art in China Mainland. Yin Yi is one of the founders of BM SPACE.

刘亚茵(编舞)Liu Yanan (Choreography)
独立舞者、编舞,与殷漪共同创办 BM SPACE。
Independent dancer, choreographer, and, like Yin Yi, a founder of BM SPACE.

范石三(影像)Fan Shisan (Photo)
摄影师,SOMC-HOUSE 工作室创始人。
Photographer, and a founder of SOMC-HOUSE, a well-known studio.

参展事务所 EXHIBITORS

都市实践 | 大舍建筑设计事务所 | 创盟国际 | 童明工作室 | 无样建筑工作室 | 直造建筑事务所
URBANUS | Atelier Deshaus | Archi-Union Architects | TM Studio | Wuyang Architects | Naturalbuild

主办单位 SPONSOR

上海市浦东新区人民政府
People's Government of Pudong New Area of Shanghai

承办单位 ORGANIZER

上海市浦东新区规划和土地管理局
Administration of Planning & Land Resources of Pudong New Area of Shanghai

协办单位 SUPPORTER

上海浦东滨江开发建设投资有限公司
Pudong Riverside Development, Construction & Investment Co., Ltd.

地点 LOCATION

浦东新区滨江大道 4700 号(近浦电路浦明路路口)
No. 4700, Binjiang Avenue, Pudong New Area (near the crossing of Pudian Road and Puming Road)

老码头煤仓外观（陈颢摄影）
The appearance of the old pier bunker (Photo: Chen Hao)

城市印记水岸传奇
闸北区苏河湾城市更新实践
案例展
City Mark – Waterside Legend
Site Project of Zhabei District

苏州河，一座伟大城市的孕育者，苏州河古名"吴淞江"，世称上海的母亲河。上海的起源和发展，都与这一脉苏州河休戚相关。先民们依河而聚，滋生繁衍，苏州河孕育了上海早期的繁荣。上海开埠后，英国侨民在吴淞江坐船逆江而上，可达苏州，故称之为苏州河。早期的工商业就沿着苏州河两岸发展，本以农渔为主业的滨江之地上海，一步步开始走向都市化、工业化、近代化，且不断向西延伸。我国近代工商业中很多"第一"在沿岸写下。这里成为巨商富贾的聚集之地，荣华流金之所。近代海上画派的领军人物吴昌硕亦卜居在此，名流大师荟粹，堪称人文渊薮。

苏河湾位于苏州河北岸。沿河而行，可以触摸到这个历史上中国民族金融业、民族工商业的发源地的前世今生。百多年前，大宗物资运输主要依靠水路，无数怀揣着财富梦想的人，沿着苏州河来到传说中遍地黄金的上海滩。苏河湾凝聚了一部现代商业冒险的恢弘诗篇。作为民族资本集聚地和重要物资集散中心，苏河湾建起了中国银行、盐业、大陆、中国实业、浙江兴业等近 20 幢欧美风格的银行仓库，素有"黄金走廊"之称。如此众多的银行在此汇金聚银，宛若金融家必争之地，充分体现其商业地位。到 20 世纪二三十年代，苏河湾已经成为上海最繁华的工商业中心，上海总商会即在此应运而生。苏河湾地区历史底蕴深厚，融码头文化、仓库文化、民族宗教文化、商业金融文化于一体，被誉为沪上"清明上河图"，积淀了老上海众多的佳话与典故。

苏河湾带来的除了近现代工业的发迹，还有直接排进河道的污染和大量生活在底层的产业工人。20 世纪后半叶，随着全球产业结构的调整，内河水运的作用逐渐衰退，苏河湾呼唤着再生的契机。从 1998 年起，上海开始了苏州河的整治工程。2002 年，逐步恢复了昔日的水体，达到景观水域的标准。亲水岸线成为更适宜居住、休闲、观光的城市生态生活区。在上海历史上，这是史无前例的回天之举。

左图	华侨城苏河湾沿岸夜景
LEFT	Night View of Suzhou Creek, Overseas Chinese Town

进入21世纪,苏州河综合整治已取得突破性进展。上海正以更高的视角思考着这条母亲河巨大的文化价值,重新打量她所潜藏的无穷底蕴。以城市复兴的理念,让她带动城市持续繁荣,焕发蓬勃的新生机。中国最具活力的大都会、"重回世界之巅"(《纽约时报》语)的上海,正成为世界聚光灯下的焦点。当我们再次跃升世界的中心时,需要一次对城市160年历史的深情回望。这场世界级的城市复兴注定从苏州河开始。

如何复兴苏河湾历经漫长的思考与论证,大量世界级资深专业机构与学者大师的潜心研究与规划,借鉴国际名城历史文化保护、利用、开发的成功经验,参考国际上成功案例进一步挖掘苏河湾的历史文化内涵和商业价值,寻找商业开发与文化传承的突破。建成后的苏河湾将成为集人文艺术、时尚商业、高端居住、都市娱乐为一体、符合可持续开发理念的上海新地标综合体。

2010年,华侨城集团进驻苏河湾,开启了这部伟大的世界级城市复兴的序幕。华侨城·苏河湾的规划蓝图,努力保持和延续城市的历史文脉,恢复旧有城市的人文性。既完美传承传统建筑风韵,又适应现代建筑美学,将创新空间与原有的城市肌理有机结合,让人感受到当代精神诠释下的文化回归与觉醒,让那个时代的现代意识和激情,皆可在此得以恢复。华侨城·苏河湾锐意探索将旧建筑保护、公共空间营造和现代新商业理念再造有机结合,在历史文化传承与时代商业精神之间,发现新上海精英生活的新内涵、新境界。

Once named Wusong River, Suzhou River cradled Shanghai and therefore is known as the mother river of this great city. The birth and growth of Shanghai are all deeply connected to Suzhou River. Ancient residents lived and developed along this river which witnessed the preliminary prosperity of Shanghai. When Shanghai became a port connecting the world, British immigrants used to go to Suzhou by ship against the current of Wusong River. Thus the river got its name. The banks of Suzhou River cradled local industrial and commercial belts.

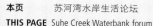

本页　　苏河湾水岸生活论坛
THIS PAGE　Suhe Creek Waterbank forum

Once a fishery-and-agriculture-focused village by the river, Shanghai then stepped on the road towards an urbanized, industrialized, modern city. Meanwhile, the city started to stretch westward continuously. Many "firsts" of Chinese modern commerce and industry were made along the river. The area was a cluster of rich merchants and fortune. Wu Changshuo, the representative of modern Shanghai Painting School, also lived here. In other words, many masters also nested in this place of rich cultural background.

Suhe Bay is located on the north bank of Suzhou River. Walking along the river, one can sense the history of the origin of Chinese finance, industry and commerce. More than 100 years ago, water transport was the main measure of transporting mass goods. Countless people came to Shanghai, a hearsay place of gold, by boat in Suzhou River, seeking for a chance to become rich. Therefore, it is safe to say that Suhe Bay is a grand poem of modern commercial adventure. As a cluster of national capital and a key entrepot, Suhe Bay is the location of nearly 20 western-style banks, such as Bank of China, Yien Yieh Commercial Bank, Continental Bank, National Industrial Bank of China and National Commercial Bank. Thus, it is known as "the golden corridor". Since so many banks, along with their capital, gather here, it is obvious that the place is the most important place for financiers and a key commercial district. Back in 1920s and 1930s, Suhe Bay became the most prosperous CBD of Shanghai and thus witnessed the birth of Shanghai Commercial Chamber. Suhe Bay features a profound history and is a place boasting rich culture related to docks, warehouses, national religions, commerce and finance. Thus, with abundant stories and anecdotes of Shanghai, it is called "Riverside Scene at Qingming Festival" of Shanghai.

Besides the prosperous modern industry, Suhe Bay also brought tremendous pollution directly to the river and numerous grassroot workers to Shanghai. In the second half of the 20th century, with the structural adjustment of the global industry, the river became a less preferred choice of transportation and thus waited a chance of rebirth. Since 1998, Shanghai has been reconstructing Suzhou River. In 2002, the river restored its previous cleanness and beauty and satisfied requirements for landscape water body. The riverside area then becomes an urban ecological place that is perfect for living, tourism and sightseeing. This is an unprecedented recovery of Shanghai.

Entering 21st Century, comprehensive regulation of Suzhou River sees a series of breakthroughs. Shanghai is weighing, from a higher view, the immerse cultural value and enormous content of this mother river, which powers Shanghai's continuous prosperity and new vitality, thus achieving urban revitalization. Shanghai, China's most energetic metropolis "back on top of the world (quoting *New York Times*) is enjoying the spotlights of the world. When we returns back to the center of the world, we need to look back at its city history of 160 years. Its rerise is destined to start here at Suzhou River.

How to restore its previous glory? It takes long-drawn-out thinking and argument, dedicated research and planning by numerous top institutions and scholars, and referring to successful practices in historical & cultural protection, utilization and development of world-famous cities. Building on these successful cases across the globe, we dig deeper the historical & cultural implication and business values, and explore breakthrough opportunity in business development and cultural inheritance. Upon completion, Suzhou Creek will rise to be Shanghai's new landmark complex of sustainable development, combining culture, are, fashion, commerce, high-end residence and urban entertainment.

In 2010, Overseas Chinese Town (OCT) announced its presence in Suzhou Creek, unveiling this great city's regeneration. OCT Suzhou Creek tries its best to conserve and continue the historical & cultural inheritance, and restores its previous humanity. On one hand, it preserves the traditional beauty of Chinese architecture, on the other hand, it assimilates modern aesthetics and combines innovation space with the existing urban texture, thus bringing cultural regression and awakening from a modern view. At here, we welcome back the modern thinking and passion of the past age. OCT Suzhou Creek explores an original approach to integrate old building protection, public space creation and modern business concepts reconstruction. Between the historical & cultural inheritance and modern business spirit we find the new meaning and frontier of elite life at Shanghai.

案例展展览主旨

70年的发展空白，为苏河湾的更新之路储蓄巨大能量与思考，当华侨城苏河湾用创新思维去考虑河流、河岸、百年建筑群、城市空间、居住空间

上图　更新展公共空间厅
ABOVE Public Space Showroom at Urban Regeneration Exhibition

下图　更新展建筑更新厅
BELOW Building Regeneration Showroom at Urban Regeneration Exhibition

上图　历史回顾墙
ABOVE Historical review wall

下图　大事件回顾墙面
BELOW Events review wall

左图　皮埃尔·于贝尔电影与录像收藏展
LEFT　Playback - Selected Works from the Pierre Huber Films and Videos Collection

之间的关系时，我们都很庆幸这个时代赋予其中的，更具文化与艺术性质的想象力，打造具有"国际性、公众性、实践性"的城市空间艺术品牌，正是上海转型之路的重要方向，也是华侨城苏河湾50万平方滨水复兴规划的核心理念。

闸北区苏河湾实践案例展以苏河湾优秀历史建筑的保护和开发实践为线索，通过不同时间轴线、不同建筑空间，将历史与建筑传承和融合，激活建筑的生命力，开展建筑、艺术、设计、人文等众多创意性的事件，创造更多鲜活的个体。在"新旧对话"的基础上，重拾每个人的城市记忆；在"传承与融合"的基础上，去挖掘创造具有新的生命力的城市空间和生活资源。

本实践案例展以"滨水城市复兴"为主题，举办一系列活动，包括三个展览：华侨城苏河湾建筑更新展、四行仓库抗战历史展（外围展）和皮埃尔·于贝尔电影与录像收藏展（外围展）；两个论坛：上海四行仓库抗战纪念地遗址保护交流研讨会和苏河湾水岸生活论坛。

展览介绍

华侨城苏河湾建筑更新展
作为本次艺术季闸北区苏河湾城市更新案例展之一的"华侨城苏河湾建筑更新展"，以华侨城苏河湾的项目开发实践为主线，重点体现对上海总商会、怡和打包厂、银行仓库群等优秀历史建筑的修复和更新，保护与传承；将视角集中于华侨城苏河湾项目规划与建设内容，整个展览从历史的回顾中展开，通过影像记录中不同人物的观点与表达、5年来的每一张图纸、在城市中看城市的体验式观感，串联起城市的过去、

235

现在和未来，就如卡尔维诺在《看不见的城市》中描写的那样，没有人能全部看懂，我们只是从一些碎片去尝试体会整个区域的转变。

修缮历史建筑立面，保留和恢复城市集体记忆；竖向绿化和城市广场，解构水岸生活层次；突出艺术和演艺空间的引领作用，创新商业模式；地下一层城市艺廊和顶层公共剧场设计，创造上海最具时尚活力的滨水体验。艺术可以为城市注入灵魂，华侨城苏河湾的践行不仅代表了城市更新，更是一次具备历史意义的开启。

四行仓库抗战历史展（外围展）
四行仓库抗战历史展是四行仓库抗战纪念馆内的永久展览。四行仓库抗战纪念馆包括序厅、"血鏖淞沪"、"坚守四行"、"孤军抗争"、"不朽丰碑"及尾厅等六个部分。纪念馆以一封谢晋元在赴淞沪战场前写给妻子凌维诚的家书开篇，展现了以谢晋元为首的"八百壮士"在国难当前之际，舍家为国的家国情怀，展现出对抗战必胜的坚定决心。展览运用实物、雕塑、现代科技等手段再现当年战斗场景，通过图文展板、巨幅绘画等形式展示上海人民投身全民族抗战、共御外侮的历史事实，以及中外各界对"八百壮士"英雄事迹的颂扬和缅怀。

回放——皮埃尔·于贝尔电影与录像收藏展
本展览由瑞士策展人圭多·斯泰格（Guido Styger）担当策划，回顾瑞士收藏家皮埃尔·于贝尔先生过去四十年间在影像艺术领域的收藏经历和成果。"回放"，是影像艺术展示过程中最为基础的状态之一。展览以此为标题，不仅出于对作品选择的整体考虑，同时也表达了对作品之后收藏家的敬意。策展人希望通过作品本身，淡化和消解主题对展览的限定，在呈现作品的同时，使观众体会到收藏行为的内在精神。

论坛及活动

上海四行仓库抗战纪念地遗址保护交流研讨会
在上海四行仓库抗战纪念地遗址保护交流研讨会上，来自全国工程技术行业的学者们以四行仓库整体修缮恢复工程为例，探讨了在上海这座中国近现代建筑遗产最多的城市中，如何在开发建筑遗产功能的同时让老建筑实现重生。

四行仓库整体修缮由曾参与修复外滩和平饭店的国家级建筑设计大师、上海建筑设计研究院资深总建筑师唐玉恩，担任整体修缮恢复工程的总设计师，由同济大学历史建筑保护实验中心对四行仓库西墙进行修缮施工。修缮过程中，工程技术人员还考虑了抗战纪念馆的功能性，在保障参观人流安全、空调通风、可持续利用等功能上下了不少工夫，开发其艺术审美和历史文化价值，让历史遗迹重生。

本页 　回放 — 皮埃尔·于贝尔电影
　　　　与录像收藏展
THIS PAGE Playback - Selected Works from
the Pierre Huber Films and Videos
Collection

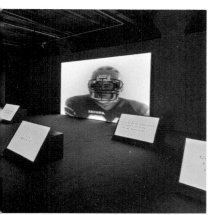

苏河湾水岸生活论坛

华侨城苏河湾"水岸生活论坛"邀请了《时代建筑》杂志的责任编辑戴春作为本次论坛的主持人,参与此项目中历史遗存更新与水岸街区生活复兴的建筑师张斌(致正建筑工作室)、邹勋(上海建筑设计研究院)、付海波(柯凯事务所)、吴迪(UNform)和艺术家刘小东,参与了研讨。各位论坛嘉宾对苏河湾项目各有着特别的渊源与情感,大家从历史建筑的保护修缮、文化传承、苏州河沿岸的区域更新等方面作了深刻的交流。无论是上海院对历史保护建筑——四行仓库的改造策略,柯凯对普通保留建筑——怡和打包厂的保护性改造,还是致正、大舍、旭可团队对五个仓库片区的研究与设计策略,以及张斌对营造一个具有复合功能的属于新上海的水岸生活空间可能性的分析,亦或是吴迪对于苏河湾建筑更新展期待呈现一个对话空间的策展出发点的诠释,均展现了嘉宾们在苏河湾这一片区的更新实践——这是一种"公共性再造",一种属于新上海生活的场域体验塑造。艺术家刘小东更以切身感受指出老建筑在给人文化依托感的同时,尺度上的适宜也让人有种亲切与舒服的感觉。

Theme of Case Exhibition

After being left unattended for 70 years, Suzhou Creek stores great energy and deep thinking for its road of regeneration. When OCT Suzhou Creek thinks, with an innovative mind, the relationship among river, banks, century-old building blocks, urban space, and living space, we are grateful of the cultural and artistic imagination created by the time. Building a urban space art brand featured by "global, public and practical" is a critical direction for Shanghai's transformation, and also the key to the restoration plan of a 500,000 m2 world-class waterside city at OCT Suzhou Creek.

With the protection and development of outstanding historical buildings in the area as links and different time axis and different architectural spaces as approaches, the Suhewan (Zhabei District) Practices and Cases Exhibition has realized the inheritance and fusion of history and architecture to activate the vitality of buildings, create more lively individuals and carry out creative events involving architecture, art, design and humanity. On the foundation of dialogue between the old and the new, it has restored everyone's memory of the city; on the basis of inheritance and fusion, it has explored the urban space and livelihood material with new vitalities.

Themed by "Waterside City Regeneration", the exhibition holds a series of events, including three exhibitions: OCT Suzhou Creek Architecture Regeneration Exhibition, Sihang Warehouse Anti-Japanese History Exhibition, and Pierre Huber Movie and Record Collections, two forums: Shanghai Sihang Warehouse Anti-Japanese Memorial Protection and Exchanged Seminar and Suzhou Creek Waterfront Living Forum.

Exhibition Introduction

OCT Suzhou Creek Building Renewal Exhibition

"OCT Suzhou Creek Building Renewal Exhibition", part of the Zhabei Suzhou Creek Urban Regeneration Projects Exhibition, focuses on project planning and construction. It starts with historical review. Different opinions and expressions recorded by video, all drawings for the past 5 years, and experience-,oriented impression from watching the city from inside, all these connect the past, the present and the future of the city. Just like Calvino says in his Invisible Cities, no one can see all things, and we just try to understand the changes from fragments. Façades of historic buildings are repaired to retain and restore the collective memory of the city; vertical greening and urban plaza construct different layers of waterside life; leading role of art and show space is highlighted for innovation in business model; design of urban art gallery at B1 and public theater at top floor create the most fashionable and vigorous riverside experience in Shanghai. Art can awaken the soul in the city, and OCT Suzhou Creek represents not only a practical practice of urban regeneration, but also a new start of historic significance.

Changes in lifestyle come even more important than urban regeneration: a city of history has the charm of time; and only a city that pursues artistic and cultural enjoyment and creation can have a better time in globalization. These are two cores of urban regeneration at OCT Suzhou Creek, and brilliant imagination for future lifestyle. This exhibition presents to the public the results of OCT Suzhou Creek for the past five years, and business & culture trend below the surface. World famous artists Yan Pei-Ming and Liu Xiaodong joined the exhibition with their latest works, showing their high praise of its ideas. The government is expected to announce, in the near future, a 100,000 m2 central park planning north to Suzhou River. OCT Suzhou Creek, a land explorer, will construct public culture space like international yacht ferry and art gallery, thus easing the life pace and enriching communication for Shanghai's core area.

The still growing Suzhou Creek has already formed strong roots in Shanghai, and is trying to reach out to the world. At here, history is conserved, and the future is created. It will be a lasting powerhouse for Shanghai's future, and bring an ideal life of global view for residents here.

Sihang Warehouse War Scenes Exhibition (peripheral exhibition)

Sihang Warehouse War Scenes Exhibition is the permanent exhibition in the Sihang Warehouse Memorial. The Sihang Warehouse Memorial includes six sections: Lobby, Songhu Battle, Defendence of Sihang, Eight Hundred Warriors, Immortal Milestone and Last Room. The Memorial begins with a letter written to the wife Ling Weicheng by the husband Xie Jinyuan before he went to the Songhu battlefield, which displays the spirit of "sacrifice for others" of the "800 warriors" led by Xie Jinyuan in the face of national crisis and their determination to win. The exhibition uses tools such as material objects, sculptures and modern science and technology to reproduce the fight scenes back then; it shows the historical facts of Shanghai people's participation in the war against aggression by means of display boards and huge paintings; people from all walks of life worldwide praise and worship for the 800 warriors' heroic deeds.

Playback-Selected Works from the Pierre Huber Films and Videos Collection

This exhibition is planned by Swiss curator Guido Styger; it reviews the Swiss collector Pierre Huber's collecting experience and achievements in the field of video art for the past forty years. "Playback" is one of the most basic states in the process of video art show. The exhibition

chooses it as the title for the sake of overall consideration of work selection as well as respect for the collector. The curator hopes to weaken and dispel the limitation brought by the subject through the works themselves; he expects that the audience can realize the inner spirit of collection behavior while presenting the works.

Forum & Events

Protection of Shanghai Sihang Warehouse Anti-Japanese War Memorial Site Exchange Seminar

On the Protection of Shanghai Sihang Warehouse Anti-Japanese War Memorial Site Exchange Seminar, scholars from the engineering industry nationwide took the overall repair and restoration project of Sihang Warehouse as an example to discuss how to renew old buildings while developing the architectural heritage function in Shanghai, the city with the largest number of modern Chinese architectural heritage.

Tang Yuen, the national-grade architectural design master and the senior chief architect of Shanghai Architectural Design and Research Institute was the chief designer of the overall renovation of Sihang Warehouse; the Historical Architecture Conservation Experimental Center of Tongji University was responsible for the renovation of the west wall of Sihang Warehouse. During the renovation, engineers have also considered the functionality of the Memorial and spared no efforts to ensure the safety of visitors, the air-conditioning and mechanical ventilation and the sustainable utilization. All efforts were made to develop the aesthetic and historical and cultural value and renew the historical site.

Suhewan Water Bank Forum

With Dai Chun, the production editor of *Time + Architecture*, being the host, "OCT Suhe Creek Water Bank Life Forum" attracted those who had participated in the renewal of historic relics and the rejuvenation of waterside communities of the Project, including Architect Zhang Bin (Atelier Z+), Architect Zou Xun (SIADR), Architect Fu Haibo (Kokaistudios), Architect Wu Di (UNform) and Artist Liu Xiaodong. All having special history and spiritual connection with OCT Suhe Creek Project, these guests exchanged profound ideas about the protection and repair of historical structures, the inheritance and promotion of culture, the upgrade of communities along Suzhou River, etc. From SIADR's strategy of reconstructing "Sihang Warehouse" (a historical building in the protection list), from Kokaistudios's protective reconstruction of Yihe Packing Factory (a non-historical building in the protection list), from the joint research and design of five warehousing areas by Atelier Z+, Atelier Deshaus and Atelier Xuk, from Zhang Bin's feasibility analysis of multi-functional waterside communities for new Shanghai and from Wu Di's communication-oriented arrangement of the Demonstration Project of Suhe Creek Architectural Renewal, people could witness these people's efforts to reconstruct Suhe Creek. It was "the reconstruction for the public". It was the establishment of field experience for new Shanghai. Based on his own experience, Liu Xiaodong also pointed out that while being cultural ballasts, old buildings were also amiable and comfortable to people if they were constructed on proper scale.

策展人感想

思班机构

华侨城苏河湾城市更新案例展围绕"城市印记，水岸传奇"的主题思想，在苏州河畔沿河商铺展出，这里是怡和打包厂的原址，开发后保留并裸露原有结构，由于新功能更新所换下的个别柱体放在了户外的草坪成为了公共艺术装置，同时把开幕式和海上明月旗袍秀放在了可以让更多市民参与的户外。展览根据空间自然分割为六个空间体验不同的版块，包括序厅、回顾墙、影像厅、建筑更新案例厅、艺术装置区和人物厅。事实上这个展览经历过建筑之外的设计展、苏河湾销售中心、华侨城成就展、规划建筑展、跨界艺术展、休闲体验文艺空间等不同定位的思考，最终以现在这种融合历史、规划、建筑、影像、装置、舞蹈、服装、音乐的复合形态呈现。

Note from the Curator

s.p.a.m

OCT Suzhou Creek Urban Regeneration Project Exhibition, themed by "City Impression & Waterside Legend", selects stores along the Suzhou River, which is the former site of Yihe Packaging Factory. The original structures are retained and exposed in the air. Some pillars left for function update are placed in the outdoor lawn to serve as public art installation. At same time, the opening ceremony and cheongsam show are held in the open air to allow more citizens to join. The exhibition is divided into six spaces naturally: preface hall, review wall, video & picture hall, building renewal project hall, art installation zone, and famous figure hall. In fact, we have various positionings for it, like design exhibition, Suzhou Creek Sales Center, OCT Achievement Exhibition, Planning & Construction Exhibition, Cross Art Exhibition, Leisure & Experience Art Space. But in the end, we select this composite manner integrating history, planning, building, video & picture, installation, dance, costume, and music.

策展人 CURATOR

姚凯 Yao Kai
上海市静安区副区长
Deputy Chief of Jing'an District, Shanghai

陈剑 Chen Jian
华侨城集团党委常委，华侨城股份公司董事、副总裁，华侨城当代艺术中心理事会理事长，深圳市城市规划委员会委员，何香凝美术馆副馆长
Standing CPC Committee Member of OCT Group, Director and Deputy President of OCT Holding Ltd., Council Chairman of OCAT, Member of Urban Planning Board of Shenzhen, Deputy Curator of He Xiangning Art Museum

袁静平 Yuan Jingping
华侨城（上海）置地有限公司总经理，OCAT 上海馆馆长
General Manager of Overseas Chinese Town (Shanghai) Land Co., Ltd.; Director of OCAT Shanghai

策展服务团队 PLAN EXECUTION TEAM

思班机构 UNIFORD
s.p.a.m., UNIFORD

主办单位 SPONSOR

闸北区人民政府
People's Government of Zhabei District

承办单位 UNDERTAKER

闸北区规划和土地管理局
Zhabei District Planning and Land Administration Bureau

协办单位 SUPPORTER

华侨城（上海）置地有限公司
Overseas Chinese Town (Shanghai) Land Co., Ltd.

地点 LOCATION

华侨城苏河湾规划展示中心 | OCAT 上海馆 | 四行仓库抗战历史纪念馆
OCT Suhe Creek Planning Demonstration Center | OCAT Shanghai | Sihang Warehouse Resistance War Museum

宜居空间
普陀区曹杨新村实践案例展
Livable Space
Site Project of Caoyang New Village, Putuo District

曹杨新村是上海第一个工人新村，把邻里单位的规划思想和大型居住区实践相结合，具有成熟的规划体系、完整的城市形态、齐备的公共服务配套、舒适的绿化环境。但是随着六十多年的更新演进，尤其是自20世纪90年代市场化更新开发的介入，社区公共空间品质日以下降，社会服务设施更新缓慢，呈现出多元的社会与空间矛盾，成为考察社区发展与更新的重要实践案例。

为此，同济大学建筑与城市空间研究所对曹杨新村开展了近八年的跟踪研究，近年又联合意大利威尼斯建筑大学开展以曹杨新村为实证对象的城市设计研究，重点包括社区更新、公共服务设施改造、公共开放空间与环境的更新等。借此次"上海城市空间艺术季"活动，对曹杨新村公共空间的城市更新的研究与实践作进一步的思考和宣传。

Caoyang New Village is the first workers' new village in Shanghai, which practically integrates the planning ideas of the neighboring units with the large residential area with the natural planning system, complete urban form and full public service facilities and comfortable green environment. However, with the upgrading and evolution for more than 60 years, particularly the emerging market-oriented upgrading development in the 1990s, the quality of the community public space dropped year by year with the slowdown upgrading of the social service facilities, so there appear the diversified social and space contradictions, which has become the important practical case of reviewing the community development and upgrading.

In this regard, Institute of Architecture & Urban Space of Tongji University conducted the follow-up research of Caoyang New Village for nearly 8 years. In recent years, it cooperated with University of Venice Institute of Architecture to conduct the urban design research with Caoyang New Village as the empirical object, particularly focusing on the community upgrading, renovation of the public service facilities, upgrading of the public open space and environment, etc. On the occasion of this activity of "Shanghai Urban Space Art Season", we conducted the further reflection and publicity of the research and practice of the urban regeneration of Caoyang New Village Public Space.

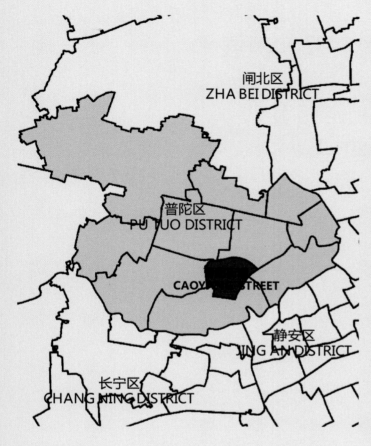

左图　曹杨新村区位图
LEFT　Location of Caoyang New Village

展览介绍

作为"上海城市空间艺术季"参展案例，曹杨新村展览以桂巷路步行街更新实践项目为切入点，结合社区文化艺术嘉年华，通过"实践与畅想"主题展览、"规划与对话"主题论坛、"空间与活动"主题演绎，来展现新中国最老的工人新村在社区环境优化、公共服务设施改造、居住品质提升等方面的城市更新探索实践与展望。

本次展览包含三部分内容：

实践案例展

配合普陀区和街道的要求，同济大学团队针对桂巷路步行商业街进行了更新改造的环境景观设计，已于 6 月底开工实施，10 月 20 日底竣工，这将成为社区公共空间改造更新的实践案例展的主体部分。展览主场地就设在桂巷路绿地广场中。

右页，上图　"亲子涂鸦"活动现场
OPPOSITE, ABOVE　Parent-child Graffiti

右页，下图　启动仪式 暨"语说曹杨"公共论坛
OPPOSITE BELOW　Launching Ceremony &"Discourse Caoyang" Public Forum

Aerial View 鸟瞰图

上图　曹阳街道文化中心方案鸟瞰图
ABOVE Aerial-view Design of the Caoyang Street Culture Center

设计研究展

结合中外院校长期的研究与设计成果，对曹杨新村未来的发展设想、结构调整、形态演进、尤其是社区环浜公共空间更新改造，提出新的构想。这些内容将以图版、模型、录像等形式展出。

纪实影像展

在曹杨新村的研究中，以摄影学和人类学的方法，记录了调查、入户访谈、设计研究、施工跟踪，将以一组"曹杨人家"录像视频，反映曹杨新村的研究、调查、访谈、设计、实施以及社会与生活状态。

上图　参观展览的市民
ABOVE Visitors

Theme of Case Exhibition

Caoyang New Village Exhibition, as the typical exhibition case for "Shanghai Urban Space Art Season", focuses on Guixiang Road Pedestrian Street upgrading practical project as the starting point by combining the Community Culture and Art Carnival to exhibition the optimization of the community environment, innovation of the public service facilities and the improvement of the residential quality to explore for the practice and expectation of the urban regeneration based on the "Practice and Expectation" theme exhibition, "planning and dialog" theme forum.

上图　参展摄影作品《路遇》(李为民)　　下图　参观展览的市民
ABOVE *Road Encounter* by Li Weimin　　**BELOW** Visitors

曹杨新村环浜绿地方案图
THE SCHEME OF PUBLIC GREENLAND AROUND THE CREEK IN CAOYANG NEW VILLAGE

曹杨新村各个小区形成封闭化管理后，原先连续贯通的环浜公共绿地被各个权属所分割，形成了断续的绿带。为此，研究小组通过调研，提出"再续环浜"的构想，打通环浜公共绿地，形成连续贯通的滨水公共活动空间，加强亲水性，提高各个社区至环浜绿地的可达性和便捷性。

With closed management of each community in Caoyang Xincun, the original continuous public greenland around the creek is segmented by ownership, which forms discontinuous greenbelts. For this, the research team has brought up the idea of "re-connecting the greenland around the creek", meant to connect public green spaces and offer a continuous public space, with more diversified waterfront activities and higher accessibility from nearby communities.

曹杨街道新步行系统方案图
THE SCHEME OF CAOYANG SUB-DISTRICT NEW PEDESTRIAN SYSTEM

研究小组为打破"门禁社区"所造成的隔阂与不便,仔细研究了各个社区历史脉络与路径,提出打造"街区渗透"的概念,从原有的封闭社区中,在保持社区安全格局下,提炼构建新的步行系统,链接至环浜绿地、公共汽车站店或公共服务中心。加强街区渗透性,有助于居民日常生活的便利。

In order to solve the problem of inconvenience of "access-controlled communities", the research team, after careful studies on their history and development, has brought up the concept of "neighborhood penetration". Without impacting safety of communities, a new pedestrian system will be designed among the existing communities, which leads to the waterfront green land, bus stops and the public service center. Integration with nearby communities is thus strengthened and the daily life will be more convenient.

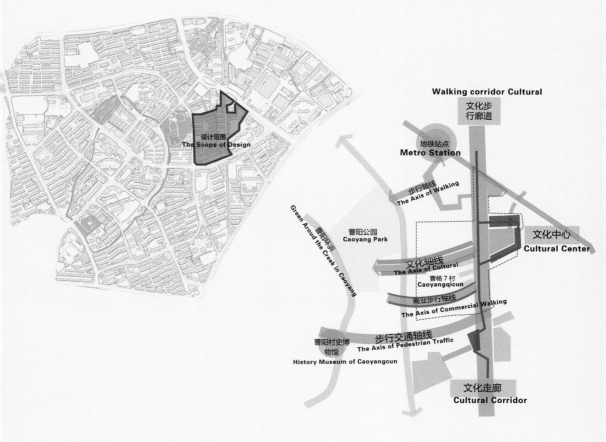

左图	曹杨文化中心设计范围
LEFT	Scope of design for Caoyang Cultural Center

右图	曹杨街道文化中心结构设计
RIGHT	Structure of Caoyang Cultural Center

Exhibition Introduction

The exhibition consists of three parts,

1. the practice of case development

Responding to the requirements of Putuo District and the Sub-district Office, the team of Tongji University conducted the environment landscape design for the upgrading and renovation of Guixiang Road Pedestrian Street. The engineering construction got started at the end of June and the project was completed on October 20, which will be the main part of the practical case exhibition for the community public space renovation and upgrading. The main venue of the exhibition was on Guixiang Road Greenland Plaza.

2. Design Research and Development

We put forward the new proposals for the upgrading and renovation of Caoyang New Village, particularly the Community Huanbang Public Space in combination with the long-term research and design results of the colleges and universities at home and abroad and in view of the future development framework, structural adjustment and form evolution, all of which will be displayed in the form of plates, models and videos.

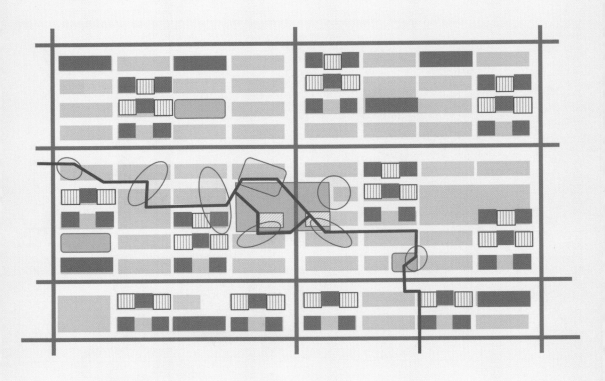

图例 LEGEND

- 织布性建筑 DARNING BUILDING
- 拆除建筑 DELETE
- 服务性建筑 SERVICE BUILDING
- 公共空间 PUBLIC SPACE
- 10-14 层建筑 10-14 BUILDING
- 空中连廊 CONNECTING LINE
- 文化建筑 CULTURE BUILDING

3. Documentary Film Festival

The photography and anthropology methods are used to record the whole process from the survey, door-to-door interview, design research and construction follow-up in the research of Caoyang New Village to reflect the research, survey, interview, design and implementation of Caoyang New Village project and the social life conditions through a group of videos.

夜跑跑道 Running Way

铺地 Ground Cover

改造建筑 Reconstruction Buildings

绿化植物 Plants

滨水阶梯 Stairs toward River

水面 Loop River

上图　曹杨街道环浜设计图
ABOVE Design drawing of Caoyang Streets by the River

策展人感想

城市空间：参与的艺术与艺术的参与

在纽约，菲利普·约翰逊设计的电报电话大楼曾经是后现代主义的象征。在它被卖给索尼公司以后，原来对公众开放的区域已经转为索尼体验店外加巨大的广告屏，索尼说这是为了公众的需要。这在纽约全城引发了"索尼广场：公共空间还是公司形象"的大讨论。

在上海，浦江两岸以震旦大厦为先导，出现了众多楼宇幕墙式广告，其硕大闪烁虽也惹得市民颇有微词，但人单势薄毕竟没用，光污染沿浦江两岸，这个城市最大的公共空间，呈不断蔓延之势。

虽说两个城市的民众在公共文化和表达意愿上存在着差别，但对于上海，在坐望全球城市目标之时，如何提高城市品质、推进城市更新，深化城市治理？在涉及城市公共价值领域等热点议题上，只有推动市民有效的参与，才会取得最广泛的成效。

2015年的秋季，上海举办了第一届"城市空间艺术季"大型公共艺术事件，来彰显城市发展进入新的转型，从过去地域的扩张，转向文化内涵的建设。而这一届的主题"城市更新"，更是点出了以公共艺术为媒介，构建专业与业余、专家与市民、艺术与生活之间的桥梁，成为推动市民认识、了解、参与自己家乡建设的盛大事件。展览中不仅重视介绍重点地区的发展和研究，更有对于民生和日常生活的深切关注，从陆家嘴到曹杨新村、从密斯范德罗奖到乡土建筑营造，期间穿插的各种报告、讲座、表演、运动等多样化的艺术的参与，让广大市民以行动表现出参与的艺术。

Note from the Curator

Urban space: Participating Art and Art Participation

In New York, the Telegraph and Telephone Building designed by Philip Johnson used to represent the post-modernism symbol. After it was sold to Sony Corporation, the original public open area was transformed into the huge advertising screen of Sony Experience Store. According to Sony, it was to meet the demand of the general public, which triggered off a big debate over "Sony Plaza: Public Space or Corporate Image" in whole New York.

There appeared various building curtain wall screen adverts in Shanghai with Aurola Tower as the leader on both banks of the Huangpu River. The large flashing advertising provoked a shocked reaction from the local citizens who didn't have the power discourse due to their weak position, but the light pollution on both banks of the Huangpu River takes a spreading trend to the largest public space in this city.

Although there is a far cry between the citizens in the two cities in terms of public culture and the expression of their respective desires, when Shanghai reviews the city targets in the world, such problems come up as to how to improve the urban quality, promote the urban regeneration and deepen urban governance. In view of the hot issues in the urban public value fields, only when the citizens are strongly promoted for the active participation can we obtain the most extensive results.

In autumn of 2015, Shanghai had the large public art event- the first "Urban Space Art Season" to highlight the new transformation of the urban development from the previous field expansion to the cultural connotation construction. However, the theme for this session "urban regeneration" represented the grand event as the public art media to build a bridge between the professionals and amateurs, experts and citizens, art and life for the grand event to promote the citizens' awareness and understanding and participation in their hometown construction. The exhibition not only emphasized the introduction of the research of the development of key areas, but also expressed the deep concern over the ordinary public and their daily life, particularly their participation, from Lujiazui to Caoyang New Village, from Mies van der Rohe Award to the local architecture creation with all kinds of reports, lectures, performances and sports activities with the diversified participation of the general public so that they can participate in the art activities with their specific actions.

策展人 CURATOR
王伟强
同济大学建筑与城市规划学院教授
Wang Weiqiang
Professor of College of Architecture and Urban Planning, Tongji University

策展团队 TEAM
同济大学建筑与城市规划学院 | 意大利威尼斯建筑大学 | 棠弥投资 有限公司
College of Architecture and Urban Planning, Tongji University | University of Venice Institute of Architecture | Tangmi Investment Co., Ltd.

主办单位 SPONSOR
普陀区人民政府
People's Government of Putuo District

承办单位 UNDERTAKER
普陀区规划和土地管理局
Putuo District Planning and Land Management Bureau

协办单位 SUPPORTER
普陀区曹杨新村街道办事处
Putuo District, Caoyang New Village Sub-street Office

地点 LOCATION
上海市普陀区曹杨新村桂巷路步行街
Guixiang Road Pedestrian Street. Caoyang New Village, Putuo District, Shanghai City

曹杨街道航拍图
Aerial map of Caoyang Streets

寻·回
上楼下乡实践案例展
Seek · Return
Site Project of Rooftop Country Experience

上楼下乡——"寻·回"艺术展，天空菜园利用具有"海绵城市"功能的V-BOX生态种植箱、生态种植基质、生态营养肥等标准化菜园种植设备，在商场、学校、企业、家庭的屋顶、阳台等空闲空间进行艺术化打造，建设绿色菜园，将新型"海绵城市"带到人们身边。

At the "Seek · Return" Art Exhibition of Going Upstairs and Coming to the Countryside, V-ROOF uses standardized garden plant equipment like V-BOX ecological plantation box, plantation base material and nutrition fertilizers equipped with the function of "Sponge City" to create artistically in free space of roofs and balconies of shopping malls, schools, enterprises and families as a way to build green gardens and bring the new "Sponge City" to the people.

展览通过场景打造，让参观者了解城市的起源，分享城市存在的问题，让参观者认识到科普打造海绵城市的重要性，参与体验全新田园生活方式，思考人类追求美好生活的本源。将城市屋顶空间利用与生活方式"更新"相结合，探索"文化兴市，艺术建城"理念的具体实践。

整个展区分为打造海绵城市工具展示区、海绵城市科普讲堂、V-ROOF【天空菜园】打造"海绵城市"案例展示区、"海绵城市"儿童实践区、"海绵城市"家庭菜园实践区、"海绵城市"屋顶菜园实践区六个区域。展览主要以场景打造、展品展示、讲解、体验的形式向市民展示。

打造海绵城市工具展示区
天空菜园利用具有"海绵城市"功能的 V-BOX 生态种植箱、生态种植基质、生态营养肥等标准化菜园种植设备，在商场、学校、企业、家庭的屋顶、阳台等空闲空间进行艺术化打造，建设绿色菜园，将新型"海绵城市"带到人们身边，为都市人提供田园休闲平台。这一区域向市民展示打造屋顶菜园的工具。让人们了解 V-BOX 生态种植箱将雨水收集重新利用的原理。

海绵城市科普讲堂
以文字、图片的形式向市民们展示人类历史发展、时代变迁的过程。从中揭示都市化发展带来的弊端——"雨后看海"，引出打造"海绵城市"的重要性，让人们了解更多"海绵城市"相关知识及打造的新思路。

V-ROOF【天空菜园】打造"海绵城市"案例展示区
以图片形式向市民展示 V-ROOF【天空菜园】近些年来所打造的屋顶农庄、企业菜园、学校菜园、私家菜园等实践案例。

"海绵城市"儿童实践区
让孩子们亲近自然、释放天性，在自然清新的环境下进行娱乐，体验种植，了解知识，让孩子们在寓教于乐的环境中成长。

"海绵城市"家庭菜园实践区
将"海绵城市"的理念结合景观艺术手法，同时融入环保技术打造私家菜园，种植蔬菜、花卉，把田园送进家庭，让人们"在都市亲手摘种蔬果"的田园梦想成为现实。在这个区域休闲让市民感受家庭田园的惬意。

"海绵城市"屋顶菜园实践区
以商业屋顶平台为载体，将田园搬入城市，为都市人构建"上楼下乡"休闲新体验，为居民提供一个种植体验平台。通过种植与都市居民共同打造"海绵城市"。

LEFT Plantation box of roof gardens

At the exhibition, we enable visitors to understand the origin of cities, share problems in cities with them and inform them of how important sponge city is through different scenes. Visitors can be part of the new pastoral life and explore the source of the pursuit of a wonderful life for mankind. Moreover, we combine the utilization of urban roof space with the renewal of way of life and explore the concrete practice of "Rejuvenating the city through culture and building it through art".

The whole exhibition area is composed of six sections: Display of tools used in building the sponge city, Sponge city popularization hall, Display of cases on how V-ROOF creates sponge city, Sponge city children practice area, Sponge city family garden practice area and Sponge city roof garden practice area. This exhibition presents items to the citizens mainly in the form of scene creation, exhibits display, interpretation and experience.

Display of Tools used in Building the Sponge City

V-ROOF uses standardized garden plant equipment like V-BOX ecological plantation box, plantation base material and nutrition fertilizers equipped with the function of "Sponge City" to create artistically in free space of roofs and balconies of shopping malls, schools, enterprises and homes as a way to build green gardens, bring the new "Sponge City" to the people and provide city dwellers with pastoral leisure platforms. Tools of building roof gardens are shown to the citizens. Visitors can know the principles of how the V-BOX ecological plantation box collects and reuse rain water.

Sponge City Popularization Hall

We show urban dwellers how humanity has developed and time has changed through words and pictures. As a result, shortcomings of urbanization are revealed. People will see the whole like a sea when a rain befalls. Thus, the importance of creating a "sponge city" is proposed. Visitors can learn more about sponge city and embrace new ideas.

Display of Cases on How V-ROOF Creates Sponge City

In this section, we inform urban residence of roof farms, enterprise gardens, school gardens, private gardens and other practical cases V-ROOF has accomplished in recent years through pictures.

Sponge City Children Practice Area

Children can get close to Nature and unleash their instincts. In a natural and fresh environment, they will have fun, feel plantation and learn knowledge. As such, children will grow and develop through edutainment.

Sponge City Family Garden Practice Area

In this area, we integrate the concept of sponge city with landscape art approaches. Meanwhile, we build private gardens, plant vegetables and flowers in an environment-friendly manner as a way to bring gardens to families and make it possible for people to plant and pick vegetables and fruits in cities. While having fun in this area, residence will enjoy the pleasure of what family gardens have to offer.

Sponge City Roof Garden Practice Area

With business roof platform as the carrier, we bring gardens to cities, thus creating a new leisure experience of Going Upstairs and Coming to the Countryside for urban people and providing a platform to feel plantation for the residence. Based on plantation, we are in a joint effort to build the sponge city with residence.

上图　前言墙
ABOVE Preface

263

策展人感想

通过这次举办上楼下乡"寻·回"艺术展,能够让更多的市民了解绿色屋顶对城市建设的重要性,进一步理解"海绵城市",也让更多的市民参与到田园种植中,助力打造"海绵城市",我觉得意义非常重大,只有全民齐动员,才能更好地改善我们的城市,建设我们的家园。

Note from the Curator

Thanks to this "Exploration and Return" Art Exhibition of Going Upstairs and Coming to the Countryside, more citizens will realize the importance of green roofs to the urban construction and further understand sponge city. Alongside that, more urban dwellers will be active players in rural planting and contribute to the efforts to build the sponge city. In my mind, such programs are of great significance. We can turn our city for the better and build our common home only when the whole population join hands to take actions.

策展人 CURATOR
庄少武
清境(上海)农业发展有限公司总经理
Zhuang Shaowu
General manager of Qingjing (Shanghai) Agricultural Development Co., Ltd.

主办单位 SPONSOR
闵行区人民政府
People's Government of Minhang District

承办单位 UNDERTAKER
闵行区规划和土地管理局
Minhang Planning and Land Administration Bureau

协办单位 SUPPORTER
清境(上海)农业发展有限公司
Qingjing (Shanghai) Agricultural Development Co., Ltd.

地点 LOCATION
闵行区七莘路 3655 号 2,凯德七宝购物广场 4 楼 V-ROOF【天空菜园】屋顶农庄
V-ROOF Roof Farm, 4th Floor, CapitaLand Qibao Shopping Mall, Building 2, No. 3655, Qixin Rd, Minhang Dist.

"天空菜园"
V-roof

上海城市规划展示馆系列展
实践案例展
Serial Site Projects in Shanghai Urban Planning Exhibition Center

一座城市的规划馆承载着传承城市历史、寄寓城市未来的重要使命。上海城市规划展示馆这座中国首家以展示城市规划和建设成就为主题的专业性场馆，2000年开馆之初即确立了"城市、人、环境、发展"的展示主题，在"解读城市历程、展望未来愿景、传播规划知识、促进公众参与"的目标定位下，16年来共接待了本地市民及海内外游客近700万人次，是上海城市形象的重要展示窗口。

A city's urban planning center should bear the responsibility of history inheritance and future outlook. As the first professional center in China based on exhibitions of urban planning and construction achievement, SUPEC, since its opening in 2000, has set the exhibition theme of "city, human, environment and development", and targeted "city's history, future vision, planning knowledge and public participation". And the center has witnessed 7 million visits home and abroad during 16 years, serving as the important exhibition window of Shanghai.

展览主旨

此次"文化传承、梦想起航——上海城市规划展示馆实践案例系列展"以上海城市空间艺术季的"文化兴市、艺术建城"活动理念为引领,围绕"城市文化遗产""城市更新实践""艺术感悟求索""少儿梦想起航"等主题展览,结合"思辨创新——论坛(讲座)""规划科普——校园展教活动""娃娃画上海——儿童广场创意绘画秀"等社会公众参与活动,在9月至12月长达4个月的系列展览活动期间共接待了9万余人次的观众,其中学生参展人数近1.5万人,并受到了7所中小学的进校园展教活动的邀请,受众学生人数达7000人次以上。

上图 上海城市规划展示馆外景
ABOVE Exterior of Shanghai Urban Planning Exhibition Center (SUPEC)

展览介绍

"上海石库门文化之旅"主题展览

"上海石库门文化之旅"主题展览由上海城市规划展示馆、上海石库门文化研究中心和同济大学建筑与城市规划学院共同主办,于9月8日至10月9日在上海城市规划展示馆二楼临时展厅展出。

石库门作为上海独有的建筑形态,近百年来与上海的城市气质和人文精神血脉相连;石库门孵化了这座城市的生活形态、滋养了独特的海派文化、孕育了大批杰出人物;石库门留下的丰富遗产,至今仍被当代的人们幸福地享用。

本次展览是上海近年来对"石库门文化遗产保护与传承"较为全面、系统的一次解读。展览由"艺术之魅、历史回眸、城中遗韵、此在彼在、留住乡愁"五个版块组成。从历史角度、文化角度、学术角度、遗产保护角度,深入浅出地解析了石库门的前世今生,演绎了石库门的文化风情,体现了石库门浓厚的上海乡情。由此,引领观众全方位地展开石库门的文化体验之旅。

"上海石库门文化之旅"是一个"解析建筑历史的学术展",展览详实地展示了石库门的起源、演进、类型、布局、发展、建筑结构、生活形态。

"上海石库门文化之旅"是一个"文学与建筑交融的实践展",展览以文学大师(张爱玲、王安忆、夏衍、矛盾等)名著中关于石库门场景的描写为原型,以当下上海石库门的现场调查为基础,由同济大学建筑与城市规划学院师生将这两个时空创造性地结合,形成近40个石库门建筑设计模型,在穿越时空而来的经典文字中凝固曾经的石库门风情,展开惊艳的跨界对话。

"上海石库门文化之旅"是一个"艺术与乡愁交织的文化展",展览展出

本页 "上海石库门文化之旅"主题展
THIS PAGE Theme exhibition of Cultural Tour of Shanghai Shikumen

上图　李守白剪纸作品　　　下图　贺友直绘画作品
ABOVE Paper-cuts from Li Shoubai　　　**BELOW** Drawings from He Youzhi

左图 石库门弄堂的生活百态 (场景模型)
LEFT Daily life in Shikumen (model)

了贺友直、李守白、慕容引刀等知名海派文化大家的主题绘画、剪纸作品,从艺术层面生动展现上海石库门的文化内涵。

"上海石库门文化之旅"是一个"可体验与聆听的科普展",展览设有多种形式丰富的公众参与活动,有石库门建筑拓片互动区、文化阅览区、公众留言区、文创产品区、石库门小讲堂、大师讲座等。

"上海石库门文化之旅"是一个"可延伸与复制的社会宣教展",作为展览的延伸,规划馆以核心展示内容为基础,将其转化成更具通俗性和科普性的展示内容,进入中小学进行展示和教育,使上海石库门文化进一步融入社会,扎根校园,为上海石库门文化遗产的传承与保护从"娃娃抓起"提供实践与推广的经验。

"回眸·定格"摄影展

"回眸·定格"摄影展由上海城市规划展示馆、上海石库门文化研究中心共同主办,于9月8日至10月9日在上海城市规划展示馆五楼临时展厅展出。

由于上海城市人口的迅速膨胀,石库门里弄开始变得非常拥挤,居住环境质量不断下降,人们开始纷纷搬离昔日的家园。近年来随着上海城市的快速发展,石库门里弄被大量拆除。它们正在逐渐地从上海人的

下图 石库门建筑立面
BELOW Elevation of Shikumen buildings

1	3
2	4

1 展厅内的阅览区
　Reading area in the exhibition hall

2 展厅现场
　Exhibition hall

3 观众在体验石库门头拓片
　Rubbings making by the public

4 展厅现场
　Exhibition hall

左图　《甘肃路153弄》摄影：乐建成
LEFT　*Lane 153, Gansu Rd.* Photo by Le Jiancheng

生活中，从城市的版图中消失。

"回眸·定格"摄影展选取了摄影师娄承浩、乐建成、寿幼森的60幅摄影作品，围绕着"消逝的石库门"这一主题，为已经逝去在时光中的石库门展现一面记忆墙。展览通过"美好记忆、窘迫现实、期待新生、留住乡愁"四个版块，使观众既感受到城市的日新月异为蜗居在石库门中的居民开启了新生活的幸福，也体会到部分具有遗产价值的建筑与生活形态在逐步消逝的遗憾，以此唤起公众对石库门文化遗产保护与更新的关注和思考。

"迁想录"—— 韩秉华艺术展

"迁想录"——韩秉华艺术展由上海城市规划展示馆、上海市闸北区文化局、景德镇陶瓷学院共同主办，于2015年10月1日至11月15日在上海城市规划展示馆三楼临时展厅展出。

"迁想录"——韩秉华艺术展主题出自东晋画家顾恺之在画论中引出的创作真谛——"迁想才有妙得"；意为艺术家只有以丰富的学养和文化积淀为基础，并在拓展创作想象空间的同时把自身的思想情感"迁移"入作品中，与对象融合，才能达到主客观的统一，才能有"妙得"的作品。展览汇聚了韩秉华先生创作的水墨、水彩、油画、陶瓷、琉璃、景观雕塑、平面设计、图象设计等艺术作品，涵盖了韩秉华先生四十多年来各时期的艺术演变及思潮。这些不同艺术领域、不同创作意境的作品，不仅体现着传统与现代的交融，更蕴含着艺术家的中国传统哲学思想对现代艺术创新的影响力。为支持首届上海城市空间艺术季的举办，韩秉华先生新近创作的以城市规划为意境的雕塑作品"成规"，与以上海建筑景观为创意的水墨作品《浦江晨曲》均为首次向公众展出。

上图　（由左至右）郑祖安、刘魁立、夏丽萍、顾骏、王慧敏、蔡丰明、张雪敏
ABOVE (From left to right) Zheng Zu'an, Liu Kuili, Xia Liping, Gu Jun, Wang Huimin, Cai Fengming, and Zhang Xuemin

下图　郑祖安研究员与刘魁立副主任在论坛现场交流观点
BELOW Exchanges between Zheng Zu'an and Liu Kuili

"娃娃画上海"——创意儿童画大赛优胜作品展暨广场创意绘画秀活动成果展

"娃娃画上海"——创意儿童画大赛优胜作品展暨广场创意绘画秀活动成果展由上海城市规划展示馆和上海市青少年活动中心共同主办，于2015年11月15日至12月27日在上海城市规划展示馆五楼临时展厅和一楼大厅展出。

展览以"发现上海 梦想上海"为主题，以4~11岁的少年儿童为参与对象。小朋友们将生活中看到的、听到的、接触到的、感受到的上海城市的美，以儿童特有的想象力、创造力通过绘画创作的形式描绘出来。活动共收到了1000余幅作品，经专家评选，最终遴选出了其中的60幅作品在规划馆展出。这些作品充满了童趣童真，所描绘的城市空间具有儿童特有的想象力和创造力。通过展览活动的举办，在引导孩子们走近城市、感受生活、表达心愿的同时，也让成人朋友们关注儿童对城市发展的期待与梦想；关注为下一代创造更为安全、美丽、文明的城市；

使城市空间艺术成为具有生命力、具有传承性、具有可持续发展的品质和前景。

上图 "迁想录"——韩秉华艺术展展厅现场
ABOVE Exhibition hall — Han Binghua Art Exhibition-Divergent Thinking Record

下图 韩秉华青铜雕塑《天马》
BELOW Bronze sculpture *Flying Horse* by Han Binghua

论坛

"2015上海石库门的保护与传承高峰论坛"

由上海市政协文史委、上海市规划和国土资源管理局主办,上海城市规划展示馆、上海石库门文化研究中心承办的"2015上海石库门的保护与传承高峰论坛",于2015年9月30日在上海城市规划展示馆举行。市、区相关部门和市高校、学术研究机构、文化遗产保护、城市文化研究、文化产业界的专家学者以及媒体记者等社会各界人士约两百人出席了论坛。

论坛主讲嘉宾上海历史学会熊月之会长,以石库门与近代上海社会的关系为切入点,分析了石库门居民的属性,不同时期的居住感受和当代人的石库门情感,从人文角度提出了城市更新必须关注城市人文特性的观点。论坛主讲嘉宾上海创意设计中心首席策划大师邵隆图先生,从塑造海派文化品牌和引领时尚经典的角度,探析了石库门文化传承与创新将在城市更新中发挥的重要作用。

论坛还邀请了文化部国家非物质文化遗产专家委员会副主任、中国民俗学会名誉理事长刘魁立先生,上海市城市规划设计研究院总规划师夏丽萍女士,上海大学上海社会发展研究中心副主任顾骏先生,上海社科院部门经济研究所创意经济研究室主任王慧敏女士,上海社会科学院文学研究所研究员蔡丰明先生,上海石库门文化研究中心主任张雪敏先生等六位对话嘉宾对上海石库门的保护与传承发表了各自观点。

论坛由上海社会科学院历史研究所研究员郑祖安主持,论坛现场气氛热烈,台上台下互动交流频繁,各路嘉宾观点鲜明,精彩纷呈,为石库门文化的保护与传承,提出了高屋建瓴的真知灼见。

论坛与会的社会各界人士一致认为,石库门是上海城市文化的名片和上海人的精神家园,也是上海城市更新发展和建设国际大都市的重要资源。保护和传承石库门文化遗产功在当代,利在千秋。

Exhibition Theme

This exhibition, led by the concept of "to prosper cities with culture and art" in Shanghai Urban Space Art Fair, focusing on the themes of "urban culture heritage", "urban renewal practice", "art perception", "children's dreams", etc., and combining "innovation of critical

thinking-forum (chair)", "popularization of science-school exhibition", "children drawing Shanghai-creative drawing in children's square", and other public activities, had been visited by 90 thousand tourists from Sep. to Dec. in 2015, among which 15 thousand were students. And 7 middle and primary schools with almost 7 thousand students were invited to enjoy the exhibition.

Exhibition Introduction

Theme exhibition of Cultural Tour of Shanghai Shikumen

This exhibition was cohosted by SUPEC, Shanghai Shikumen Culture Study Center, and College of Architecture and Urban Planning of Tongji University in the temporary exhibition room, located at the 2nd floor of SUPEC, with an exhibition period from Sep. 8 to Oct. 9.

Shikumen, as a unique architectural form only found in Shanghai, has been weaved into the urban culture and humanistic spirit of Shanghai; it breeds Shanghai's life style, nourishes Shanghai-style culture and cultivates numerous talents; it provides abundant heritage which has been enjoyed by people today.

Through this exhibition, an overall and systematic understanding on "cultural heritage protection and inheritance" was made by Shanghai in recent years. It consists of "the charm of art, historical review, city's traditional charm, well-balanced past and future, and lingering homesickness". And in this exhibition, Shikumen was analyzed in a simple way from history, culture, academy, and heritage protection and its cultural aroma with strong provincialism of Shanghai was deduced, making the tourists fully immersed into the Shikumen cultural tour.

It was an academic exhibition analyzing architectural history, in which detailed origin, evolution, types, layouts, development, architectural structures and living forms of Shikumen were displayed.

It was a practice exhibition blending literature and architecture, in which the prototypes of Shikumen in masterworks by literature masters (Eileen Chang, Wang Anyi, Xia Yan, Mao Dun, etc.) were adopted; site research was made about the current Shanghai Shikumen; and the two were creatively combined by teachers and students in College of Architecture and Urban Planning of Tongji University. Almost 40 models of Shikumen architecture were built, and they realized the dialogue between the past Shikumen in literatures and the present Shikumen in front of us.

It was a cultural exhibition blending art and homesickness, in which theme drawings and paper-cuts from famous Shanghai-style cultural masters, such as He Youzhi, Li Shoubai, Murong Yindao, were displayed. It vividly revealed cultural meaning of Shanghai Shikumen.

It was an audiovisual science-popularizing exhibition, in which there were many forms of activities that the public could participate in, such as rubbing making of Shikumen architecture, culture reading, the public's message-leaving, cultural creation products, lecture room for Shikumen and master's chair.

It is an extensible and duplicate exhibition with social propaganda and education. Based on the key exhibitions, the Center turned them into something with more science popularity, and displayed it in middle and primary schools, which would promote the influence of Shikumen culture in schools; therefore, the children would know the importance of inheritance and protection of Shikumen culture; this served as good experience for practice and propaganda.

Photographic exhibition of Glancing Back-Freeze Frame

This exhibition was cohosted by SUPEC and Shanghai Shikumen Culture Study Center in the temporary exhibition room, located at the 5th floor of SUPEC, with an exhibition period from Sep. 8 to Oct.9.

Due to the fast expansion of urban population in Shanghai, Shikumen lanes became very crowded, and living environment there deteriorated, so people started to move. In recent years, as rapid development occurs in Shanghai, lots of lanes have been torn down. They are disappearing from the city and from people's view.

In this exhibition, 60 pictures was chosen from photographers, including Lou Chenghao, Le Jiancheng and Shou Yousen, focusing on the theme of "disappearing Shikumen" to form a memory wall full of old Shikumen. The wall consisted of four parts, respectively standing for "beautiful memory, awkward reality, outlook of new lives and lingering homesickness". It would make the residents still in Shikumen experience both the happiness brought by the quick change of the city and the regret brought by the disappearing architecture and living forms with heritage value, in order to arouse the public's attention to the protection of Shikumen cultural heritage.

Han Binghua Art Exhibition-Divergent Thinking Record

This exhibition was cohosted by SUPEC, Shanghai Culture Bureau (Zhabei), and Ceramic Institutes of Jingdezhen in the temporary exhibition room, located at the 3rd floor of SUPEC, with an exhibition period from Oct. 1 to Nov.15, 2015.

The theme of the exhibition originates from the creation essence-"Divergent Thinking Brings Wonderful Gains" in the theory of painting by Gu Kaizhi, a painter in the Eastern Jin Dynasty; it means that only the knowledgeable and cultivated artist with cultural accumulation would gain wonderful things when he migrates his emotion into the works during the expansion of imaginary space. In this exhibition, Han's artistic works were displayed, such as Chinese ink painting, watercolor, oil painting, ceramics, colored glaze, landscape sculpture, plane designs and image designs, which covered over 40 years of evolution and ideological trend of Han. These creative works from different artistic fields not only represented the traditional and modern blending, but also contained the influence of Han's traditional philosophy over modern artistic creation. In order to support the first art fair, Han would firstly show his new sculpture work-"Chenggui" based on artistic conception of urban planning, and his Chinese ink painting-"Morning Tune on Pujiang River" based on Shanghai's construction landscape.

Exhibition of Children's Creative Drawing Competition-"Children drawing Shanghai"-Creative Drawing Show in Children's Square

This exhibition was cohosted by SUPEC and Shanghai Youth Activity Center, located at the 5th floor of SUPEC and the ground floor, with an exhibition period from Nov.15 to Dec.27.

The exhibition took "discoveries and dreams in Shanghai" as its theme, and aimed at children

of 4 to 11 years old. The works were drawn with children's imagination and creativity as well as from their view of Shanghai's beauty. 60 pictures were chosen by experts from 1000 ones collected in the activity, and were displayed in the Center. These works were filled with children's simplicity. This activity was helpful to make the children closer to the city and life, and for them to convey their own willingness; it also helped adults pay attention to children's expectation on urban development, as well as to building a more beautiful, civilized and safer city for them. In sum, it would make the urban space art become vigorous, continuous and sustainable.

上图　儿童绘画作品展示
ABOVE Children's drawings

下图　儿童广场创意绘画秀
BELOW Creative drawings show in children's square

Forums

2015 Summit Forum for Protection and Inheritance of Shanghai Shikumen

2015 Summit Forum for Protection and Inheritance of Shanghai Shikumen, sponsored by Shanghai Political Consultative Culture and History Committee and Shanghai Planning and Land Resource Administration Bureau) , undertaken by SUPEC and Shanghai Shikumen Culture Study Center, was hosted in SUPEC on Sep. 30, 2015. About 200 people from local authorities, colleges, academic organizations, cultural heritage protections, urban culture studies, cultural industries as well as medias, attended this forum.

Xiong Yuezhi, chairman of Shanghai History Academy and guest speaker of the forum, based on the relation between Shikumen and recent Shanghai, analyzed features of residents in Shikumen, living experiences in different ages, and Shikumen complex of modern people, and proposed that urban renewal should focus on caring for humane characteristic. Shao Longtu, another guest speaker for the forum and chief planner in Shanghai Creative Design Center, based on Shanghai-style culture brand and leading fashion, discussed the importance of Shikumen culture inheritance and innovation to the urban renewal.

Liu Kuili, vice director of National Intangible Cultural Heritage Experts Committee and honorary president of China Folklore Society, Xia Liping, chief planner of Shanghai Urban Plan and Design Research Institute, Gu Jun, vice director of Shanghai Social Development Research Center of Shanghai University, Wang Huimin, director of Creative Economy Office of Institute of Economics in Shanghai Academy of Social Sciences, Cai Fengming, research fellow of Institute of Literature Study in Shanghai Academy of Social Sciences, and Zhang Xuemin, director of Shanghai Shikumen Culture Study Center were invited as guests in the forum to express their views about how to protect and inherit Shikumen culture.

The forum was hosted by Zheng Zu'an, a research fellow of History Study Institute of Shanghai Academy of Social Sciences.And the forum atmosphere was hot with frequent interactions; guests from all circles expressed their clear and brilliant views which would serve well the protection and inheritance of Shikumen culture.

All of the attendants agreed that Shikumen should be on behalf of Shanghai's urban culture and the spiritual home of people in Shanghai, and that it should be the important resource with which urban renewal and international metropolitan would be done. Protection and inheritance of Shikumen cultural heritage would benefit future generations.

策展人感想

最初提出参与首届城市空间艺术季的初衷只是渴望作为以城市规划为主题的专业性场馆能为城市文化活动尽一份力。但对"文化兴市、艺术建城"理念的深入思考，激起了我对"城市文化的根基与个性、艺术创新的环境与自省、城市更新的目的与追求"进行探究的欲望。所以才有了石库门的遗产价值、艺术家的"迁想"情怀、孩子们眼中的上海，这三个既是独立题材又紧密关联的系列展览的构思。不论是城市的"精神"、艺术家的"个性"还是孩子的"天性"，都源自于"天赋"、成就了"眼界"、又受益（或受制）于环境，但更关键的是能否在把自身的情感赋之于"作品"的同时（不论这个"作品"是城市、是物体、是他人还是自我），去敬畏自然规律、敬畏传统文化、敬畏利益法则，敬畏"作品"的尊严和生命力。所以上海城市空间艺术季的举办不仅代表着这个城市当下的活力与追求，更预示着这个城市未来的品质与成就。是城市创新思维由单一走向多元、由限制走向包容、由桎梏走向开放，由参与走向众创的催化剂。一旦城市的规划者、建设者、管理者、参与者能真正感悟到上海城市空间艺术季活动的内涵，从而为城市的未来、为居民的愿景、乃至为自我生存价值的回归留出更多的思考与创新空间的话，那么艺术与梦想的求索一定能使我们的城市迸发出无限的可能。

Note from the Curator

That I proposed to participate in the first Shanghai Urban Space Art Season was to make the professional pavilion based on urban planning helpful to urban cultural activities. By deeply understanding the concept of "to prosper cities with culture and art", I desired to explore "the root and character of urban culture, environment of artistic innovation and self-examination, as well as aim and pursuit of urban renewal". Therefore, I conceived the exhibition with three independent but connected themes, including the value of Shikumen heritage, the artists' divergent thinking and the Shanghai city in children's eyes. The city's "spirit", the artist's "individuality" and the child's "instinct" stem from "talent" and "mind", and benefit (are restricted by)the environment. It is more important that while bestowing emotion on "works" (city, object, others or ego), we would show our respect to natural law, traditional culture, benefit law as well as vitality and dignity of the works. So, the art fair not only represents current vigor and pursuit of Shanghai, but also predicts the quality and achievement of the city. It serves as the catalyst which turns unitary creative though into diversified ones, restriction into inclusion, shackle into openness, and participation into public creation. Once the planners, builders, managers and participants of the city comprehend the essence of the art fair and are willing to ponder the city's future, the residents' vision, and the return self-value, the pursuit of art and dreams is bound to make our city burst out of the infinite possibility.

总策展人 GENERAL CURATOR
翁文斌 Weng Wenbin
上海城市规划展示馆总工程师
Chief engineer of Shanghai urban planning exhibition hall

《上海石库门文化之旅》主题展 策展人
CURATOR OF "THEME EXHIBITION OF CULTURE TOUR OF SHANGHAI SHIKUMEN"
翁文斌 Weng Wenbin
上海城市规划展示馆总工程师
Chief Engineer of Shanghai Urban Planning Exhibition Hall

张雪敏 Zhang Xuemin
上海石库门文化研究中心主任
Director of the center for Shanghai Shikumen Culture

威广平 Qi Guangping
同济大学建筑与城市规划学院副教授
Associate Professor, Institute of Architecture and Urban Planning, Tongji University

《回眸·定格》摄影展策展人
CURATOR OF "PHOTOGRAPHIC EXHIBITION OF GLANCING BACK-FREEZE FRAME"
翁文斌 Weng WenBin
上海城市规划展示馆总工程师
Chief Engineer of Shanghai Urban Planning Exhibition Hall

张雪敏 Zhang Xuemin
上海石库门文化研究中心主任
Director of the center for Shanghai Shikumen Culture

特别致谢 SPECIAL THANKS TO
阮仪三教授 Ruan Yisan
同济大学国家历史文化名城研究中心主任
Professor, Director of National historical and cultural city research center ,Tongji University

韩秉华先生 Han Binghua
香港美术家协会副主席
Vice Chairman of Hong Kong Artists Association

张建龙教授及其师生团队 Prof. Zhang Jianlong & his team
同济大学建筑与城市规划学院
College of Architecture and Urban Planning ,Tongji University

张岱虹女士及其工作团队 Zhang Daihong & her team
上海市青少年活动中心
Shanghai Youth Activity Center

贺友直先生 He Youzhi
中国美术奖·终身成就奖获得者 连环画泰斗
Winner of Lifetime Achievement Award, China Fine Arts Prize;The high priest of Comic Strip

李守白先生 Li Shoubai
上海海派剪纸艺术大师
Shanghai Paper-cut Art Master

慕容引刀先生 Murong Yindao
当代漫画家"刀刀狗之父"
Contemporary Cartoonist, the Father of "knife knife dog"

地点 LOCATION
上海城市规划展示馆（人民大道 100 号）
Shanghai Urban Planning Exhibition Center (No. 100, Renmin Avenue)

上海中心城核心区域大模型
Shanghai central city core area model

致谢

上海市规划和国土资源管理局
上海市文化广播影视管理局
上海市徐汇区人民政府

经过一年多的筹备，2015上海城市空间艺术季案例展在策展团队、国内外学者，以及社会各界的共同努力下成功举办了。来自世界各地的规划师、建筑师和艺术家们围绕"城市更新"主题，展示了各自多年来的研究成果，为上海的城市建设、城市更新工作提出了独到的见解，同时也为上海市民搭建了一个了解城市更新历史、感受上海城市文化的互动平台。展览期间，社会各界对此次展览的内容、形式，以及布展方式等均给予了高度的评价，同时也提出了很多有建设性的意见和建议。此次展览的举办为组织下一届上海城市空间艺术季积累了宝贵的工作经验。

展览在举办过程中得到了多方的大力支持和帮助，谨代表2015上海城市空间艺术季的承办单位，特别鸣谢以下单位和团体：

浦东新区人民政府、黄浦区人民政府、静安区人民政府、长宁区人民政府、普陀区人民政府、闸北区人民政府（原）、虹口区人民政府、杨浦区人民政府、宝山区人民政府、闵行区人民政府、嘉定区人民政府、金山区人民政府、松江区人民政府、青浦区人民政府、奉贤区人民政府、崇明县人民政府、上海市城市规划设计研究院、上海西岸开发（集团）有限公司、上海市测绘院。

另外，我们也要感谢对艺术季曾经给予支持和付出努力的领导、专家、策展人、工作人员和热情参与的市民。

本画册中部分资料和图片由原创作者免费提供，在此一并表示感谢！

Acknowledgements

Shanghai Municipal Bureau of Planning and Land Resources

Shanghai Municipal Administration of Culture, Radio, Film & TV

People's Government of Xuhui District, Shanghai

After more than a year of preparation, The Site Project of Shanghai Urban Space Art Season 2015 was successfully held with the joint efforts of the curatorial team, domestic and international scholars and people from all walks of life. At the SUSAS themed by "Urban Regeneration", planners, architects and artists from all over the world displayed their achievements of many years' research, offering unique insights into urban construction and regeneration in Shanghai and creating for Shanghai residents an interactive platform to understand the history of urban regeneration and feeling the urban culture of Shanghai. During the exhibition, people from all sectors of society spoke highly of the content, form and layout methods of the show while providing many constructive comments and suggestions. From this exhibition, we have gained valuable experience for the next Shanghai Urban Space Art Season.

Considering that we have gained vigorous support and assistance from multiple parties during the exhibition, we hereby, on behalf of the organizer of Shanghai Urban Space Art Season 2015, would like to express our special gratitude to the following entities and organizations:

People's Government of Pudong New Area, People's Government of Huangpu District, People's Government of Jing'an District, People's Government of Changning District, People's Government of Putuo District, People's Government of Zhabei District (former), People's Government of Hongkou District, People's Government of Yangpu District, People's Government of Baoshan District, People's Government of Minhang District, People's Government of Jiading District, People's Government of Jinshan District, People's Government of Songjiang District, People's Government of Qingpu District, People's Government of Fengxian District, People's Government of Chongming County, Shanghai Urban Planning and Design Research Institute, Shanghai West Bund Development (Group) Co. Ltd. and Shanghai Municipal Institute of Surveying and Mapping.

Meanwhile, we would like to appreciate those people who have given support and made efforts on SUSAS including experts, curators, staff, citizens and their enthusiastic participation. In addition, some data and pictures in this album are provided by courtesy of the authorship. Thank you for all the support!

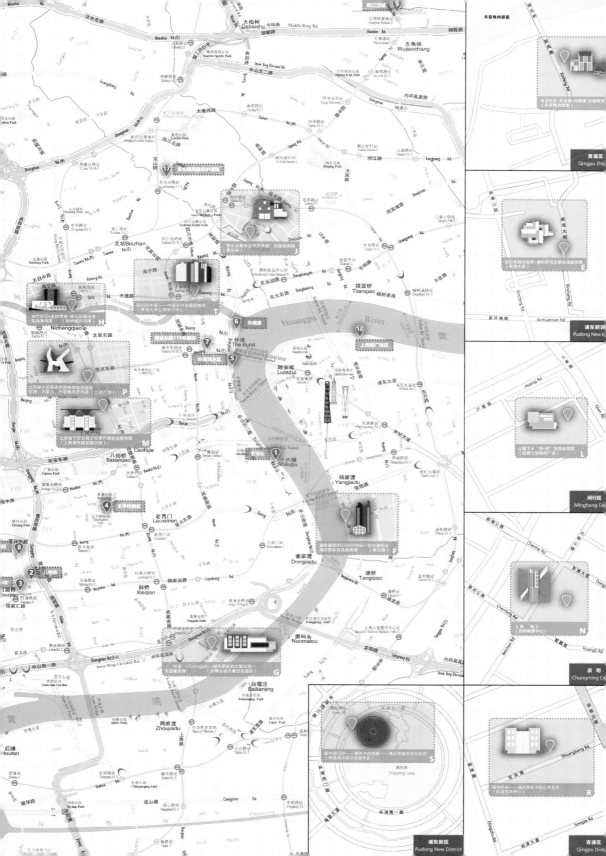

图书在版编目（CIP）数据

2015 上海城市空间艺术季案例展 = 2015 Shanghai Urban Space Art Season Site Project : 汉英对照 / 上海城市空间艺术季展览画册编委会编 . -- 上海 : 同济大学出版社 , 2016.7

ISBN 978-7-5608-6296-5

Ⅰ. ① 2... Ⅱ. ①上 ... Ⅲ. ①城市规划一空间规划一上海市一画册 Ⅳ. ① TU984.251-64

中国版本图书馆 CIP 数据核字 (2016) 第 083904 号

2015 上海城市空间艺术季案例展
2015 Shanghai Urban Space Art Season
Site Project

上海城市空间艺术季展览画册编委会 编
Edited by SUSAS Publication Editorial Board

策划：上海城市公共空间设计促进中心 群岛工作室
项目统筹：马宏
责任编辑：杨碧琼
特约编辑：邹野
责任校对：徐逢乔
装帧设计：绵延工作室 | Atelier Mio
版 次：2016 年 7 月第 1 版
印 次：2016 年 7 月第 1 次印刷
印 刷：上海安兴汇东纸业有限公司
开 本：787mm x 1092mm 1/16
印 张：18.5
字 数：370 000
书 号：ISBN 978-7-5608-6296-5
定 价：178.00 元
出版发行：同济大学出版社
地 址：上海市四平路 1239 号
邮政编码：200092
网 址：http://www.tongjipress.com.cn
经 销：全国各地新华书店

本书若有印装质量问题，请向本社发行部调换。
版权所有 侵权必究